USE OF SUGARS
AND OTHER CARBOHYDRATES
IN THE FOOD INDUSTRY

A collection of papers comprising
the Symposium on the Use of Sugars
and Other Carbohydrates in the Food
Industry, presented jointly before the
Divisions of Agricultural and Food
Chemistry and Carbohydrate Chemistry at
the 123rd meeting of the American
Chemical Society, Los Angeles,
Calif., March 1953.

Number 12 of the Advances in Chemistry Series
Edited by the staff of *Industrial and Engineering Chemistry*

Published February 1955, by
AMERICAN CHEMICAL SOCIETY
1155 Sixteenth Street, N.W.
Washington, D. C.

Copyright 1955 by
AMERICAN CHEMICAL SOCIETY

All Rights Reserved

CONTENTS

Introduction

SIDNEY M. CANTOR

Sidney M. Cantor Associates, Ardmore, Pa.

The papers appearing in this volume were selected from a symposium sponsored by the Division of Carbohydrate Chemistry with the cooperation of the Division of Agricultural and Food Chemistry.

In making the original selection of topics to be covered, specialists were invited to discuss specific carbohydrates, uses of carbohydrates in particular industries, and nutritional and regulative aspects. This was done in order to provide several points of view. The participants were asked also to consider their subjects in terms common to the chemist and therefore not obscured by their particular specialized terminology. The reason for this was the recognition that the choice of a carbohydrate constituent or constituents to be incorporated in a foodstuff is made by the manufacturer not only on the basis of sweetness as, for example, in the case of sucrose, but for a substantial number of other physical, chemical, and economic reasons as well.

The large variety of carbohydrate constituents available to the food manufacturer offers a broad spectrum of properties when used individually or in combinations which range from high to low with respect to sweetness, viscosity, consistency, solubility, and many others. Furthermore, the proper manipulation of these properties by the variety of processes and techniques at the disposal of the food chemist magnifies the multiple role which the carbohydrate constituents play.

Thus, there can be added to the taste and appearance qualities which the carbohydrate constituents contribute, their use as preservatives in canned foods, in the manufacture of preserves and jellies, and in the curing of meats, their use as carbon dioxide sources in the baking industry, and in a host of other applications, as these papers testify.

The discussion of these properties in physical and chemical terms provides a common denominator among the food technologies and allows developments in one segment to contribute to other segments. It should help to dispel the popularly held notion that sugars are associated only with sweetness and starches only with thickening.

The contents of this volume then are dedicated to a better understanding of the ways in which our largest single dietary constituent—the carbohydrates—contributes to the physical and chemical nature as well as the nutritional quality and acceptability of our foods.

The Role of Sugar in the Food Industry

ROBERT H. COTTON[1], PAUL A. REBERS, J. E. MAUDRU, and GUY RORABAUGH

Research Department, Holly Sugar Corp., Colorado Springs, Colo.

The role of sugar in the food industry is described here in terms of its chemical and physical properties. Studies of the chemical aspects of sugar include those on sweetness and flavor, the reactions of sucrose and invert sugar, caramel formation, antioxidant effect, and the effect of sugar on the curdling of milk, gel formation, and metal corrosion. Physical properties of sugar discussed here are osmotic pressure, crystallization and solubility, hygroscopicity, thermodynamic properties, viscosity, grain size, bulk handling, and such miscellaneous properties as stickiness and thermal and electrical conductivity.

Sugar plays a major role in the production of thousands of food products from cured meats through preserves and frozen fruits to confections. Sugar (sucrose) is white, colorless crystalline compound whose molecule is composed of one molecule each of D-glucose (dextrose) and D-fructose (levulose) with the elimination of one molecule of water. Its organic chemical name is 1-D-glucopyranose β-D-fructofuranoside. Part of the extraordinary versatility of sugar lies in the fact that it can be hydrolyzed partially or completely to the two simple sugars, dextrose and levulose.

Wartime scarcities of sugar served to prove dramatically how basic is sugar to this country's food supply. Nutritionally, sugar produces energy. Economically, sugar is probably the most efficient foodstuff in terms of calories produced per acre tilled (43). With the great advances in nutrition over the last 50 years it may be perhaps easily forgotten that adequate calories are essential to health (66). Hockett (43) has shown that after one meets the minimum daily requirements for protein, fat, minerals, and vitamins, as recommended by the National Research Council, there still remains approximately 1500 calories of the total 3000 considered essential to a 150-pound man. These 1500 calories can be supplied by any wholesome food of choice. Since sucrose contributes palatability to many foodstuffs rich in proteins, minerals, or vitamins, it also serves nutritionally in addition to producing calories. As an example, addition of sucrose to processed orange juice rich in vitamin C but with a low sugar-acid ratio makes the juice palatable and thus results in utilization of a valuable food which otherwise would be wasted. Furthermore, sugar enhances the stability of canned orange juice (93) and many other foods. Thus, it would seem that rational use of sugar is both nutritionally and economically desirable. The extent to which sugar is used is shown in Table I.

With the growth of food science or technology many of the "arts" of food preparation and preservation yield to scientific inquiry and method. More and more chemists and engineers are entering food processing industries (105). These industries will more and more use sugar in light of its basic chemical and physical properties to produce a given result and in accord with carefully controlled tests on quality and yield. The latter approach requires both great effort and expense as it involves carefully conducted taste panels and expert food technologists working in well equipped laboratories. It may involve extensive market testing. Such work fits into the competitive American scheme where emphasis on quality often means the difference between profit and loss. Such work also leads to new products (105).

---

[1] Present address, Huron Milling Co., Harbor Beach, Mich.

The growth of the rational use of sugar calls for widespread knowledge of the chemical and physical properties of sucrose. In this paper an attempt is made to indicate some of the sources of knowledge in this field and to give a few examples of newer food technology studies on sugar.

Table I. Sugar Deliveries, by Type of Product or Business of Buyer
(United States, year ending December 31, 1950[a])

Product or Business of Buyer	Total Sugar, 100-Lb. Units
Bakery, cereal, and allied products	12,723,819
Confectionery and related products	14,376,515
Ice cream and dairy products	5,074,003
Beverages	15,115,929
Canned, bottled, frozen foods, jams, jellies, preserves	11,009,726
Multiple and all other food uses	4,993,283
Nonfood products	780,019
Hotels, restaurants, institutions	554,051
Wholesale grocers, jobbers, sugar dealers	60,775,357
Retail grocers, chain stores, supermarkets	23,440,775
All other deliveries, including deliveries to government agencies	1,758,747
Total deliveries	150,602,224

[a] Represents over 95% of deliveries by primary distributors in continental United States. Data from Production and Marketing Administration, USDA.

Chemical Aspects of Sugar

Sweetness and Flavor. Because the taste sensation is a subjective phenomenon, indexes of relative sweetness must always represent averages of opinion. In spite of this difficulty Dahlberg and Penczek (27) and Cameron (13, 14), using improved testing methods, have been able to obtain reproducible results. Both groups have shown that the relative sweetness of the various sugars varies with the concentration, as is shown in the curve in Figure 1. The results of Dahlberg include values for corn sirup and corn sirup solids (Table II).

Table II. Iso-Sweet Sugar Solutions According to Dahlberg (27)
(Concentrations in per cent by weight)

	Solutions Equivalent to	
	10% sucrose	20% sucrose
Sucrose	10	20
Dextrose	12.7	21.8
Levulose	8.7	16.7
Maltose	21.1	34.2
Lactose	20.7	33.3
Enzyme	...	...
Converted corn sirup	17.9	28.2

The relative sweetness of mixtures of sucrose and other sugars, as compared to sucrose, has been studied by Dahlberg (27) and Cameron (13, 14) (see Figure 1). Dahlberg found that a solution containing 10% sucrose and 5.3% glucose was equal in sweetness to a solution containing 15% sucrose. The flavor-enhancing power of sucrose has also been noted by other workers (16).

The relative sweetness of invert sugar as compared to sucrose has been studied by Cameron (13, 14) and Miller (74). At concentrations of 10% they are equivalent in sweetness, at concentrations below 10% sucrose is sweeter, while above 10% invert is sweeter. However, because of the sweetness-enhancing power of sucrose, a solution of partly inverted sucrose will be sweeter than one completely inverted.

The type of sweetness varies with the different sugars (27). The sweetness of sucrose is quickly perceived and promptly reaches a maximum intensity, whereas the sweetness of dextrose stimulates the taste organs more slowly and reaches a maximum intensity later. The primary taste of glucose (14) is sweet but the secondary tastes are bitter, sour, or tart, while in the case of sucrose secondary flavors are absent.

The effect of temperature on sweetness in comparing the relative sweetness of sucrose to levulose has been made by Yamazaki et al. (128). Comparing 5 and

10% solutions of each sugar, they found that when they had been kept at temperatures below 50° C. levulose was sweeter, at 50° C. the sweetness was equal, and above 50° C. sucrose was sweeter. They explained the relative change of sweetness on the relative proportions of fructose isomers present at the various temperatures. At lower temperatures more of the sweeter beta isomers are present.

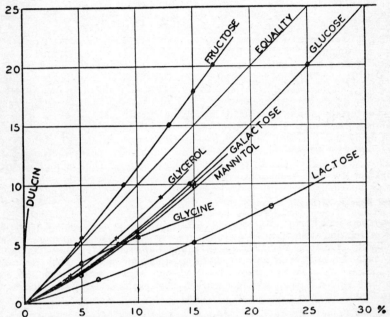

SUCROSE %

Figure 1.   Data of Cameron and Dahlberg (14, 27) on Relative Sweetness of Various Compounds Compared to Sucrose

Abscissa represents grams of compound per 100 grams of solution. Curve labeled "equality" represents sucrose—for example, a 15% lactose solution is as sweet as a 5% sucrose solution

In further discussing the relation of configuration to sweetness Yamazaki *et al.* (*128*) and Cameron (*14*) report that α-glucose and α-galactose are sweeter than their beta isomers. Yamazaki points out that in the three cases mentioned the sweetest form of each sugar possesses a cis arrangement of hydroxyl groups between the carbonyl carbon and the adjacent carbon. Although sucrose does not possess any free hydroxyl groups on the carbonyl carbon, it appears that the sweetest isomer of fructose is joined to the sweetest isomer of glucose through a glycosidic bond.

In addition to imparting the property of sweetness, sucrose is able to augment the body of flavor even when sweetness is not noticeably increased. This seasoning property of sucrose has been studied by Caul (*16*) in the cooking of corn, peas, tomatoes, and many other vegetables. It was noted that sugar and salt blend while mutually decreasing saltiness and sweetness. This property is of particular importance in those foods that contain added salt as a preservative, such as cured meat.

The amount of sucrose in a cake formula affects the texture (*63*). It is adjusted to as high a level as permissible, too high values causing the cake to fall. A great increase in the ratio of sugar to flour in cakes made possible by the development of high emulsifying shortening was followed by a markedly greater acceptance of bakery cakes by the American public (*122*).

A highly favorable effect on color, flavor, and texture of pie crust was obtained by the addition of small amounts of sucrose to the batter (*122*).

The taste interrelationship of monosodium glutamate and sucrose has been studied by Lockhart and Stanford (62). This should be of value in research in certain processed foods.

Wilcox, Greenwood, and Galloway (124) found that meat quality and dressing percentage could be increased by feeding sugar to livestock prior to slaughter.

A partial explanation of the flavor enhancement properties of sucrose may be related to its ability to promote dissociation of weakly ionized compounds. Davey and Dippy (28) reported that the conductance of five monocarboxylic acids is greater in 20% sucrose solutions than in water. The dielectric constant, a property associated with ionization, is decreased by the addition of sucrose (67).

The relative merits of sucrose, dextrose, and corn sirup as a food preservative from the standpoint of color and flavor retention have been investigated by several workers. Using frozen peach slices (18, 113, 116), strawberry preserves (110), and frozen blackberries and raspberries (17), the following results were observed: (1) Replacement of part of the sucrose by dextrose resulted in poorer color of frozen peaches and (2) replacement of 25 to 50% of the sucrose with corn sirup produced detectable changes in flavor in all the cited foods.

Another physiological study involving sweetness of sugar suggests a possible mechanism by which certain dietary preparations work. Goetzl et al. (33) found that 20 grams of sucrose ingested by human subjects at noon, in concentrations above 20% sucrose, gave rise to typical after-meal satiety. Thus, a high-sugar food should aid in a weight reduction diet by helping one to forego, or reduce, a meal. It might well combine a high bulk ingredient, such as pectin or carboxymethylcellulose.

Reactions of Sucrose and Invert. Sucrose in neutral or very slightly alkaline solutions is very resistant to decomposition by heat. However, prolonged heating at a pH of 9 or greater will result in the formation of dark coloring matter and decomposition of the sugar (104). On the other hand, reducing sugars—for example, dextrose, levulose, or invert—are much less stable in alkali and decompose to complex products (102).

Yeast readily metabolizes sucrose, but the fructrose moiety is not so readily consumed as the glucose. This brings up an interesting question under study at the present time as to the fate of sucrose in bread baking and the effect of residual sugars on the baking losses, color, resistance to drying out, and flavor of the bread. Barham and Johnson (6) give data on the influence of various sugars on dough and bread properties. A fairly recent method of quantitative paper chromatographic analysis can be adapted to the determination of sugars in food products (29, 55).

Sucrose hydrolyzes to dextrose and levulose under acid conditions, especially at elevated temperatures. Continued exposure to low pH and high temperature results in production of a few parts per million of hydroxymethylfurfural, as well as caramel, biacetyl, and methyl glyoxal (111). Hydroxymethylfurfural has reported bacteriostatic activity (69, 79). This compound, presumably from fructose, has been found in foods which have undergone browning (85). Wolfrom et al. (126) show it to be a precursor in color formation when glucose reacts with glycine. Recent work (9, 10) indicates that hydroxymethylfurfural is less important in color production in heated sucrose solutions than other sugar breakdown products. Obviously color production is desirable in certain foods, especially in crusts on baked goods, while in other foods color development is not desired. Since sucrose does not enter into Maillard reactions (58) while invert sugar does, the food technologist can avoid or minimize inversion of sucrose where color is not wanted and, on the other hand, promote it where color is desired. Furthermore, certain desirable flavors arise from Maillard reactions (7, 63). A large quantity of milk in caramels develops a flavor that can be obtained in no other way, perhaps as a result of Maillard effects (63). A detailed discussion of the Maillard reaction is beyond the province of this paper. Essentially, however, the reaction involves the aldehyde group of a reducing sugar and a free amino group, such as in amino acids.

Hodge and Rist (44) have shown the role of the Amadori rearrangement in color formation following reaction between aldoses and amino acids. The reaction can eventually lead to the development of caramel-like flavors, melanoidin pigment, fluorescence, and an increase in reducing powers. Of possible interest here is a reaction reported by Lewis, Esselen, and Fellers (61) concerning the decarboxylation of amino acids or even simple carboxylic acids at elevated temperatures. As would be expected, the use of reducing sugars as a preservative or sweetener accelerates the Maillard reaction. In the case of lemon juice concentrates this reaction results in off flavors. Wray (127) reports that the substitution of as little as 15% sucrose by dextrose can be detected. The off flavor develops as well in all-sucrose packs but only after a longer period of storage. Vacuum-dried orange juice also (38, 95) develops typical melanoid color and flavor in storage. It contains reducing sugars and amino acids (114). One successful technique for retarding the reaction is "in-package" desiccation (22). A good review is that of Lea (58). He points out that when lactose is replaced by sucrose in dried skim milk, production of insolubility and discoloration by Maillard reaction was prevented or retarded. This is, so far, of theoretical interest only; however, the author indicates that sucrose, when added before drying, increases the storage life of dried eggs. Frozen peaches are prone to develop color in storage and thawing. Sucrose was compared to dextrose and high conversion corn sirup in producing packs of frozen peaches. Of the packs prepared with a single type of sweetening agent, the sucrose samples had the best color (113).

Sugars are important in color development in meat curing. Greenwood et al. (35) state that sugars here have the role of improving color by helping to establish reducing conditions, of tending to prevent oxidation of ferrohemoglobin to ferrihemoglobin in storage, and of helping to conserve the meat during curing by its protein sparing action. By in vitro experiments with sugars plus hemoglobin plus bacteria they conclude that reducing sugars have certain advantages over sucrose as to ready utilization by microorganisms, but they also indicated that oxyhemoglobin solutions containing dextrose at certain concentrations oxidized to the brown methemoglobin more quickly than sucrose-containing solutions.

Sucrose in concentrations from 0.5 to 2.0 parts per 100 ml. of homogenized milk frozen and stored at −23.3° F. retarded for 185 days a noticeable separation in thawed milk and slightly retarded appearance of an oxidized flavor (5). This study of Babcock et al. exemplifies the modern approach mentioned previously, in which the effects of sucrose are measured quantitatively in controlled experiments. The separation they observed has a parallel in frozen concentrated orange juice. Here sugar content in storage (more precisely, total solids) had a marked effect on the enzymatic clearing of the reconstituted juice (20). Cruess et al. (25) state that when lemon juice is brought to 50° Brix by addition of sucrose the juice does not separate even after 1.5 years' storage at 0° F. It is to be hoped that some day a systematic study will be made of the effect of sugar on enzymatic reactions in foods.

Caramel Formation. The extraordinary versatility of sucrose is again exemplified by the formation of caramel. The food processor, by the proper selection of conditions, is able to cause caramel formation when it is desired for a coloring and flavoring agent or, if so desired, is able to minimize its formation. When solid sucrose is heated, caramel formation is rapid and its composition depends upon pH, time, and temperature (132). Caramel formation can be expected from prolonged overheating of heavy sugar solutions in direct contact with heating surfaces.

According to present knowledge, sucrose caramel is a complex mixture of sugar anhydrides (100). Among them are those having the empirical formulas $C_{12}H_{20}O_{10}$, $C_{24}H_{36}O_{18}$, and $C_{36}H_{50}O_{25}$. The empirical formula of sucrose is $C_{12}H_{22}O_{11}$.

Caramel is colloidal in nature and its isoelectric point, the pH at which it coagulates and precipitates, is of great importance in those foods in which it is used for coloring and flavoring, such as beer, soft drinks, sirups, soups, pickles, candy, etc. (19). The isoelectric point is fixed with its method of manufacture and may vary from a pH of 6.9 for a typical beer caramel to a pH of less than 3 for a typical soft drink caramel. There are other caramels found outside this range. If a caramel having an isoelectric point of 4.6 were to be used in soft drinks having a pH range of 4 to 5, flocculation would occur.

Antioxidant Effect. Sucrose possesses very appreciable antioxidant properties (57, 59). This is important in color, flavor, and ascorbic acid retention.

The solubility of oxygen in sucrose solutions is less than in water. The data of Joslyn and Supplee (51), MacArthur (64), and Miller and Joslyn (72, 73) show that the decrease in solubility is proportional to the concentration of sucrose. At a temperature of 20° C. a sugar solution having 60° Brix exposed to the air contains only one sixth as much oxygen as water alone.

The inhibition of the autoxidation of ascorbic acid in the presence of sucrose, dextrose, levulose, and corn sirup has been studied by various workers (47, 50, 60, 77, 91, 96, 97, 109). The autoxidation can be reduced by 10 to 90%, depending upon the concentration of sugar, the pH, and the amount of copper present. Shamrai (97) and Joslyn and Miller (50) attribute the reduced rate of the autoxidation of ascorbic acid at pH of 4 and above to the complexing action of copper by sugars. Copper is an effective oxidation catalyst even at concentrations as low as 1 p.p.m.

Joslyn and Miller (50) further reported that the physiologically active oxidation product of ascorbic acid, dehydroascorbic acid, was preserved to a greater extent in sugar solutions than in a control in which water was substituted for the sugar solution.

The retardation of the autoxidation of ascorbic acid in food products has been reported for strawberry and black currant juices (101), jams (89), and many other foods. Apricots packed in enough sucrose sirup (preferably 40° Brix) to cover the fruit required a minimum amount of added ascorbic acid to retain good color (45). A practical example of the antioxidant properties of sugar sirups for quick frozen apple slices is given by Grab and Haynes (34). Guadagni (36) describes a process in which tissue gases are removed by vacuum and this is followed by the partial replacement of the voids with a sugar sirup containing a small amount of antioxidant. The product is superior to the old type using blanching and sulfur dioxide (36).

Another practical example showing the preservative effect of added sucrose is given by Cruess and Glazewski (24) in the preservation of frozen pack fruit nectars. Nectars made from apricots—peach, plum, and guava and various blends—can be prepared and stored at 0° F. for a long period of time with little loss of fresh fruit flavor.

The addition of sucrose retarded the development of oxidative rancidity in sunflower oil (56). Sucrose added to eggs before spray drying retarded loss of vitamin A (90) and beating power (31).

It appears evident that where quality is a paramount consideration selection of sweeteners and processing techniques should be governed by careful testing and controlled experiments.

Curdling of Milk. Pecego (86) claims that addition of 4 grams of sucrose per liter of milk with or without addition of enzyme gave a cheeselike coagulation while addition of lactose or glucose gave a gelatinous one.

Sugar in Gels. Pectic substances are treated in detail in this symposium by Joseph. The role of sugar in gel formation is considered to be primarily one of dehydration (54). Sugar and acid both affect gelatinization of starch. Steiner (105) discusses applications to food production.

Corrosion. Corrosion of aluminum cooking vessels by acid fruit juices decreases as the concentration of sucrose increases (87). Edible beet molasses and, to a lesser degree, sugar solutions have a reported antirust effect when painted on metal surfaces (115). Affermi (1) has studied the effect of boiling sucrose solutions on many metals, including nine steel alloys. Sucrose has buffering capacity (41) which may explain, in part, its retardation of corrosion.

Physical Properties of Sucrose

"Polarimetry, Saccharimetry, and Sugars," produced by Bates and fellow scientists of the National Bureau of Standards and published in 1942 (8), and Browne and Zerban's "Physical and Chemical Methods of Sugar Analysis" (12) are the greatest storehouses of valuable data in the field. In addition to analytical methods one will find extensive data on viscosity, solubility, boiling points, electrical conductivity, crystallography, and melting points of the various sugars. A few newer sources of data are indicated in the following discussion of physical properties and their use to the food technologist. A good recent source of physical data on liquid sucrose and invert has been published by Junk, Nelson, and Sherrill (53). Valuable graphical presentations are given for data on refractive indices of sugar and corn sirups, sucrose inversion rates under varying conditions, safe densities for storage of invert sugars, and viscosities of several types of liquid sugars.

Osmotic Pressure. Steiner *(105)* has recently reviewed the role of sugar in preserves. Information in the following paragraph is taken from his review.

Sugar inhibits mold and bacterial growth in foods owing to the osmotic pressure of its solutions in high concentration. In jams and jellies the concentration of soluble solids in the jelly portion should exceed 72.5% in order to prevent mold and yeast growth. Since at 20° C. sucrose dissolves only to the extent of 66.4 *(112)* parts per 100 parts of sirup, mold growth can take place. However, if sucrose and invert sugar are present in equal amounts, saturation corresponds to 75% soluble solids. Cooking of acid fruits to which sugar is added will result in inversion, of course. One should try for about 30 to 35% inversion in jams and jellies when packed. They will continue to invert in storage. Glucose or corn sirup can be added to sucrose to increase osmotic pressure. The upper limit of the sugar addition to preserves is determined by saturation above which crystallization of dextrose hydrate occurs. In the case of chocolate, sugar preserves through osmotic effect, but here crystallization is allowed to take place. However, grinding and mixing reduce the size of sugar crystals to under 0.001 inch. At this size it is impossible to detect roughness or grittiness in the product.

Tables of osmotic and activity coefficients of sugar have been published by Robinson and Stokes *(94)*. These should aid the food technologist in calculating osmotic strengths.

There are a few rare cases where increase in sugar and total soluble solids tends to encourage rather than inhibit microbiological growth, as for example, with orange juice and tomato juice *(2, 76)*.

Sugar as molasses has a powerful dehydration effect which can be used as a control tool in moisture determination by food technologists *(20)*.

Othmer and Silvis *(84)* have correlated boiling point elevations of sugar solutions as a function of pressures, concentrations, and percentage purity (see their nomograph in Figure 2). When one is producing a candy or preserve where boiling point is used to control cooking time and total soluble solids in finished product, a change in atmospheric pressure can cause a serious error. Figure 2 should help eliminate such error.

When foods destined to be frozen are treated with sucrose solutions, osmotic effects are to be expected. Guadagni *(36)* has shown how these effects vary with the concentration of sugar in sirup applied to apple slices. With a 20% sucrose solution the slices continue to gain in weight with increased "residence time" in the solution, while at 30 to 60% sucrose the slices lose weight after the first 10 minutes' residence time in sucrose, presumably because of the osmotic strength of the higher concentration of sucrose. Hohl *et al.* *(46)* give data on losses in soluble solids when water is used to blanch or cool fruits and vegetables prior to freezing.

Recently a detailed study of freezing point depression of an 8.5% sucrose solution *(4)* was published as part of a milk research program. Young, Jones, and Lewis *(131)* and Junk, Nelson, and Sherrill *(53)* discuss the entire range of sucrose concentrations. Obviously, foods are not pure sucrose solutions and thus freezing points of juices, for instance, will differ somewhat from pure sugar solutions. For example, Schroeder and Cotton *(95)* have given data on the freezing curve of orange juice, which is approximately 50% sugars on a solid basis.

Crystallization and Solubility of Sucrose. Part of the versatility of sucrose, as it imparts different properties to prepared foods, lies in its solubility and the nicety of control that can be effected on the solubility. Saturated solutions of sucrose in water contain 66.4 parts of sucrose at 20° C. per 100 parts of solution and 76.4 parts at 70° C. *(112)*. Supersaturation occurs easily and by properly controlling it, speed of crystallization and crystal size may, in turn, be controlled. Spencer and Meade *(103)* give supersaturation curves.

The candy technologist, in manufacturing the various types of product ranging from hard candy and caramel to liquid-center chocolates, employs many of the physical and chemical properties of sucrose *(71)*. In hard candies, crystallization and stickiness are the two most common problems limiting shelf life *(23)*. Use of the proper amount of an acidic "doctor," such as cream of tartar, causes enough inversion of the sucrose to eliminate grain formation. Excessive invert sugars increase hygroscopicity of the hard candy. Proper amount of inversion results in a stable hard candy which becomes neither crystalline nor tacky and which retains bright color and full flavor. Colloids such as pectin, gelatin, or albumin likewise inhibit crystallization of sucrose. In general, the addition of other substances to a sucrose solution increases solubility and inhibits crystallization (15, 26, 37) (see Figures 3 and 4). In the manufacture of liquid-center chocolates, the center must be solid for chocolate enrobing. Later enzymatic inversion of sucrose by invertase liquefies the center because of formation of invert sugar. Although candy making has, in the past, been an art, it is showing a gradual trend toward use of scientific control and research (98).

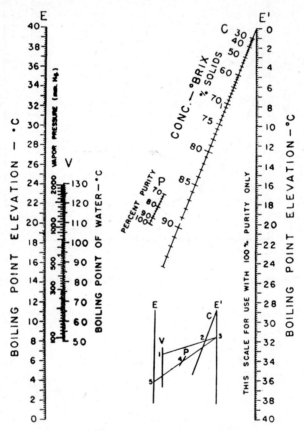

Figure 2.   Boiling Elevation of Sugar Solutions as a Function of Concentration and Purity

Taken from Othmer and Silvis (84)

With the advent of frozen foods, some of which contain added sucrose or invert sugars, a problem arose around the unsightly formation of hydrates of the sugars used, but this is a relatively rare phenomenon in commercial practice (131). Within the temperature range of —30° to +40° F., eight solid-phase sucrose hydrates have been observed (129). These hydrates appear as white spherulitic masses which may include the whole mass of frozen food after prolonged storage. Sucrose hydrates do not readily form spontaneously but usually require seeding. They melt rapidly at room temperatures. Dextrose hydrates, appearing as white beadlike masses throughout the food readily form without seed and persist even after the food has been thawed and warmed to room temperature (11). One hydrate of levulose, the hemihydrate, forms as spherulitic masses of fine needles from concentrated solutions but has, so far, not been reported in frozen foods (130). Difficulties caused by hydrate formation may be minimized by use of hermetically sealed containers during storage, careful sanitation to prevent stray seeding of hydrates, proper balance between amounts of sucrose invert sugar and/or corn sirup used, and selection of the proper storage temperature. In one study (11), minimum hydrate formation occurred at −30° F. while maximum formation was observed at −10° F.

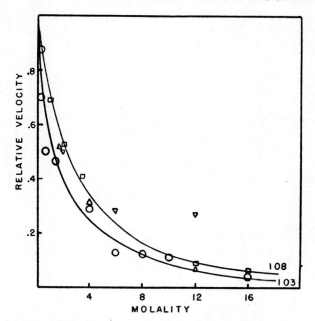

Figure 3. Effect of Invert Sugar on Crystallization Velocity
From Van Hook (117)

Sucrose plus Water		Supersaturation Coefficient	
○	0.709	1.03	
△	0.722	1.05	
▽	0.722	1.05	(stirred during adjustment)
□	0.743	1.08	

Hygroscopicity. The sugar crystal is slightly hygroscopic in its pure form (78). In addition, minute amounts of ash and a thin coating of nonsugar impurities in commercial sugar influence the humidity equilibrium values over the whole range of relative humidity. The moisture content of commercially refined sugar is very low when in equilibrium with humidities below the critical range. Average moisture in commercial sugar below the critical point is between 0.01 and 0.02%; these moistures, however, show a sharp break at exposure to atmospheric humidities over 75% (78).

Figure 5 shows the per cent moisture in sugar versus relative humidity of three different sugars at different levels of ash content.

Sugar is a free-flowing product if kept at a moisture content below 0.02% and will keep in free-flowing condition if not subjected to relative humidities above 60% for extended periods of time. Optimum storage relative humidity is 40 to 60%, which usually can be obtained by warehouse heating and humidification. When it is allowed to absorb moisture, the moisture dissolves a small amount of sugar from the crystal face until film of saturated solution of minute thickness is formed. When the sugar is then subjected to lower humidities and the moisture is dried out, this minute film recrystallizes and the points of contact of the crystals will be then fused together and caking will result.

This difficulty is more evident in sugars of small or mixed particle sizes, since more points of contact exist. Increasing the pressure on a pile of sugar increases tendency to cake.

To extremely fine sugar, such as powdered sugar, starch and/or calcium phosphate in the amount of 1 to 3% is added to prevent the sugar from caking. In addition, powdered sugar is packaged in paper bags or cartons with a moisture-vapor barrier incorporated in the container. This may be either a waxed paper or an asphalted laminated sheet.

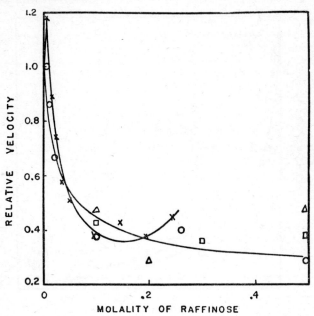

Figure 4.   Effect of Raffinose on Crystallization Velocity of Sucrose
From Van Hook *(117)* and Hungerford and Nees *(48)*

Sucrose plus Water		Coefficient of Supersaturation	Temp., ° C.	Stirring
✕	0.710	1.045	25	Gentle
○	0.722	1.05	30	None
□	0.722	1.05	30	Gentle
△	0.722	1.05	30	Rapid

The so-called soft sugars, or brown sugars, require care opposite to granulated. These sugars have a moisture content of 1.5 to 4%, and it is essential to maintain this in order to maintain the texture of the product *(78)*. Storage humidities between 60 and 70% are recommended for soft sugars whenever they are to be left in storage for long periods. Again, moisture-resistant paper is used in packaging.

In the bakery industry invert sugar is used to prevent drying and checking *(63)*.

Thermodynamic Properties. Precise design of food processing equipment and accurate specifications of process details will be facilitated by the work of Higbie *(39, 40)* on heat content (more precisely enthalpy) changes in sucrose and sucrose solutions as caused by heating, dissolving, dilution, crystallization, and concentration. This final report *(39)* gives specific examples of the use of his data and equations and will soon be published by the Sugar Research Foundation. An example of the use of the heat capacity of sucrose solutions is the technique of Noyes *(81)*. He sprays a sugar solution at about 10° to 17° F. over moving fruit as a first step in production of frozen foods. Heat transfer between the food and a liquid will be faster than when air is the cooling medium, Hohl and coworkers *(46)* have found. In the case of apple slices, Noyes *(81)* claims the sucrose-cooling technique results in improved flavor texture and yield. Spencer and Meade (*103*) give specific heat data for sucrose solutions.

Viscosity. Viscosity is a prime consideration in design of liquid handling systems for sugars. Viscosity affects texture of certain foods, such as preserves and candies. Sucrose viscosities at various temperatures and sugar concentrations determined by Landt are given by Browne and Zerban (*12*). Othmer and Silvis (*83*) have prepared a nomograph (Figure 6) correlating viscosity with temperature, concentration, and purity. Junk *et al.* (*53*) give data for pure sucrose solutions and solutions inverted 20, 50, and 90%. Calculations from their data indicate that at 40° C. a 71° Brix sirup, which is 90% inverted, has a viscosity roughly 43% of that of pure sucrose at 71° Brix. Figure 7 gives data from this laboratory on the effect of inversion on viscosity. Thus, sucrose is versatile in providing a wide range of viscosity, depending on the degree of inversion. Sucrose viscosity can be further modified by additions of other compounds.

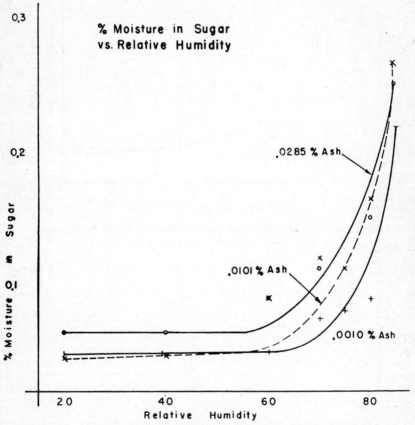

Figure 5. Moisture Content of Sugar as a Function of Relative Humidity
Taken from Maudru and Paxson (*70*)

Grain Size of Sugar. Sugar is produced and available to food processors in many grain sizes. The names of the types of different grain sizes vary in individual companies but, in general, they will fall into one of the classes in Table III. The grain size greatly influences the properties, such as ease of mixing and speed with which the sugar dissolves. Nitzsche (*80*) has found the rate of solution without agitation of white sugar in water decreases with decrease in crystal size, while the opposite is true where agitation occurs. The largest percentage of all sugar sold falls into the fine or extra-fine granulated class (65). This is an all-purpose sugar which functions satisfactorily in most of the food uses.

Methods of expressing grain size vary from individual sugar companies to individual food industries. While screen analysis is the common method of determining the grain size of sugar, types of screen vary and unless the size opening is specified for each mesh or type or standard screen used, misconception can exist. Table III gives the mean aperture (M.A.) values (88) of the different types of sugars. The ranges in mean aperture values in the table were determined by plotting the screen analysis of a number of different sugars on the market in each class. A few individual sugars in each class may fall on either side of these limits but almost all sugars will be defined by Table III.

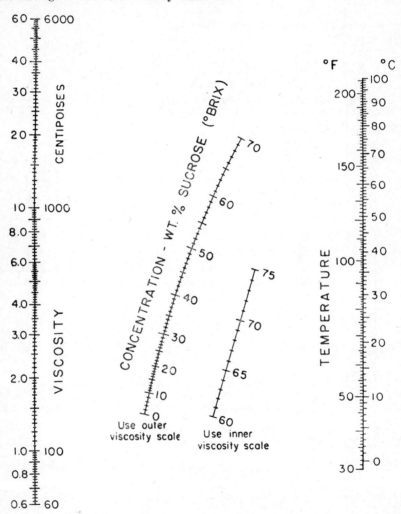

Figure 6.   Nomograph Correlating Viscosity, Temperature, and Concentration of Sucrose Solutions

Taken from Othmer and Silvis (83)

The Powers arithmetic probability plot (88) has been found by Van Hook (118) to be the most satisfactory manner of representing the size distribution of sugar crystals. Figure 8 shows three different types of sugars plotted on a screen analysis graph used for commercial control purposes: Baker's Special, extra fine, and coarse granulated. The mean aperture is taken at the point the line crosses the 50% retained line and the evenness of the grain is expressed mathematically as the coefficient of variability (C.V.). This is calculated by the formula:

$$C.V. = \frac{\text{aperture for 16\% retained } - \text{ aperture for 84\% retained}}{2 \times \text{mean aperture}} \times 100$$

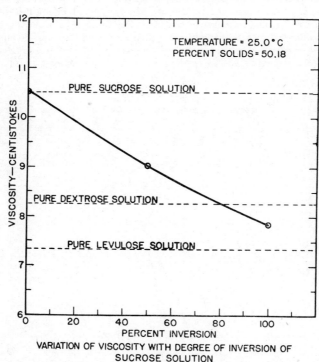

Figure 7. Variation of Viscosity with Degree of Inversion at Constant Solids Content of 50.18%

From the two values, mean aperture and coefficient of variability, a screen test may be totally defined. Greater use of the Powers probability plot would eliminate confusion by buyers and sellers alike.

Other sugars not in Table III are the brown or soft sugars, which are of fine grain mixed with cane sirups of natural color and flavor or boiled from selected low purity sirups to give a soft-textured sugar with the color and flavor desired.

Table III. Particle Size and Use of Commercial Sugar

Mean Aperture, Inch	Type	Application	Remarks
Above 0.025	Large crystals Confectioner's A Confectioner's AA Sanding	Candy manufacturers Bakeries Sanding	Used for sweetening, finishing, decorative
0.0150 to 0.0250	Coarse granulated	Jellies, sanding, confectionery, sirups, bottlers	Sweetening, jelling, preserving, dough-raising, hot-process sirup manufacturing
0.007 to 0.0150	Fine granulated Extra fine granulated	Household, baking, canning, bottlers, sirups, prepackaged food mixes	Largest percentage of all sugar sold; cold-process sirup manufacturing Sweetening gelatin mixes, cake and pudding mixes
0.002 to 0.007	Baker's Special Industrial fine Crushed fine	Baking, chocolate manufacturing Bar sugar	High solution rate and where size must be reduced by milling with other ingredients, no starch added
0.002 and below	Regrained Powdered Fondants	Dusting, cakes, icings, fillings, candies	1% tricalcium phosphate or 3% starch added to prevent caking in powdered sugar; fondants sold either wet or dry— smallest particle size sugar

These sugars are sold and packaged moist and used by bakers and confectioners, mostly as a source of color and flavor in icings, candies, and cakes.

Bulk Handling. Sucrose is versatile in its physical forms. It is sold as liquid sugar, or in packages and, finally, in bulk. Thus, sucrose can be conveyed by pump, conveyors, or even an air stream. Each system has advantages for different sizes and types of commercial installations. For example, at the Exchange Lemon plant in Corona, Calif. (*123*), bulk crystalline sugar is received in special truck trailer units (see Figure 9) and is screw-conveyed to a storage bin. From storage the sugar goes to weigh lorries traveling on an overhead monorail system and thence to process. These lorries are automatically weighed to the desired level and the weights are simultaneously printed on a tape; a record is thus kept of every batch of product.

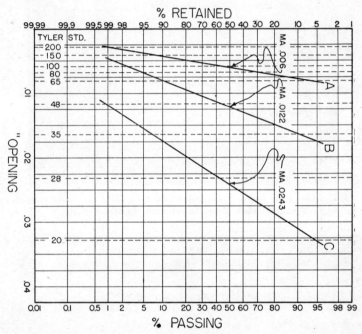

Figure 8.   Grain Sizes of Three Typical Commercial Sugars
In inches

Figure 9 shows a storage and unloading installation. Figure 10 shows the bulk handling system to the process included in this installation. An automatic 100-pound scale can be preset to make any number of weighings from 1 to 60. In addition, the rate of the desired number of weighings can be preset. This allows for addition of the sugar at the optimum rate to the solution in which it is to be dissolved. Time requirements for handling sugar were cut in half by this installation. Use of bulk sugar results in lower purchase price and plant handling costs (*30*). Bulk installations have recently been described in detail (*107, 108, 125*). Liquid sugar is discussed elsewhere in this symposium.

Miscellaneous Physical Data. Viscosity and stickiness are not identical. A method is available for measuring stickiness (*68*) which may be of interest in candy making.

Sucrose diffusion studies are reported by several authors (*42, 82, 119*) and surface tension studies are now in the literature (*12, 120*). Sherrill recently gave information on the thermal expansion of crystalline sucrose (*99*). Thermal conductivity (*92*) and electrical conductivity (*32*) of sucrose solution recently have been reported. As research in food science continues it is felt these data may be of value. Of more obvious immediate usefulness is the work of Zerban and Martin (*133, 134*) on the refractive indices of dextrose and invert sugar solutions.

Figure 9.   Bulk Sugar Transport Truck Unloading at Pacific Citrus Products' Bulk Sugar Installation, Fullerton, Calif.

Figure 10.   Interior View of Pacific Citrus Products' Bulk Sugar Installation

Note overhead automatic scale which can discharge at a variable controlled rate of either tank or the hopper in the background

Acknowledgment

The authors wish to thank James H. Turner, F. W. Blaney, Clyde Schmer, Grace Dillon, and Ed Flavell for help in assembling reference material for this paper.

Bibliography

(1) Affermi, E., *Ann. chim. appl.*, **39**, 381–92 (1949).
(2) Anderson, E. E., Esselen, W. B., Jr., and Fellers, C. R., *Food Research*, **14**, 499–510 (1949).
(3) Anderson, G. L., Jr., Higbie, H., and Stegeman, G., *J. Am. Chem. Soc.*, **72**, 3798–9 (1950).
(4) Aschaffenburg, R., and King, J., *Analyst*, **76**, 2–4 (1951).
(5) Babcock, C. J., Strobel, D. R., Yager, R. H., and Windham, E. S., *J. Dairy Sci.*, **35**, 195–8 (1952).
(6) Barham, H. N., Jr., and Johnson, J. A., *Cereal Chem.*, **28**, 463–73 (1951).
(7) Barnes, H. M., and Kaufman, C. W., *Ind. Eng. Chem.*, **39**, 1167–74 (1947).
(8) Bates, Frederick, *et al.*, Natl. Bur. Standards, *Circ.* **C-440** (1942).
(9) Beacham, H. M., and Dull, M. E., *Food Research*, **16**, 439 (1951).
(10) Bergdoll, M. S., and Holmes, E., *Ibid.*, **16**, 50–6 (1951).
(11) Brekke, J. E., and Talburt, W. F., *Food Technol.*, **4**, 9, 383 (September 1950).
(12) Browne, C. A., and Zerban, F. W., "Physical and Chemical Methods of Sugar Analysis," New York, John Wiley & Sons, 1941.
(13) Cameron, A. T., *Can. J. Research*, **23E**, 139 (1945).
(14) Cameron, A. T., Sugar Research Foundation, New York, *Sci. Rept. Ser.* No. **9** (December 1947).
(15) Casbianca, J. B., *Sucr. franç.*, **90**, No. 6, 16–18 (1949).
(16) Caul, J. F., *Sugar Molecule*, **5**, No. 2 (1951).
(17) Caul, J. F., and Sjostrom, L. B., *Quick Frozen Foods*, **13**, No. 10, 59–61, 111 (1951).
(18) Caul, J. F., Sjostrom, L. B., and Turner, W. P., *Ibid.*, **12**, No. 4, 54–8 (1950).
(19) Clark, P. T. C., *Intern. Sugar J.*, **53**, 162 (1951).
(20) Cotton, R. H., Harris, W. A., Orleans, L. P., and Rorabaugh, Guy, *Anal. Chem.*, **24**, 1498 (1952).
(21) Cotton, R. H., Roy, W. R., Brokaw, C. H., McDuff, O. R., and Schroeder, A. L., *Proc. Florida State Hort. Soc.*, **1947**, 39–50.
(22) Cotton, R. H., and Schroeder, A. L., U. S. Patent 2,520,878 (Aug. 29, 1950).
(23) Cramer, A. B., *Food Technol.*, **4**, 400–3 (October 1950).
(24) Cruess, W. V., and Glazewski, I. G. A., *Frosted Frozen Foods* (July 1946).
(25) Cruess, W. V., Seagrave-Smith, H., and Glazewski, I. G. A., *Quick Frozen Foods*, **9** (April 1947).
(26) Dahlberg, H. W., and Bennett, A. N., *Ind. Eng. Chem.*, **43**, 660 (1951).
(27) Dahlberg, A. C., and Penczek, E. S., N. Y. State Agr. Expt. Sta., *Bull.* **258**, (1941).
(28) Davey, I., and Dippy, J. F. J., *J. Chem. Soc.*, **1944**, 411.
(29) Dubois, M., Gilles, K., Hamilton, S. K., Rebers, P. A., and Smith, F., *Nature*, **168**, 167 (1951).
(30) *Food*, **21**, 30–1 (1952).
(31) Fryd, C. F. M., *Food Manuf.*, **21**, 471–5 (1946).
(32) Gardner, S. D., and Hill S., *Brit. Abstr.*, **C2357** (1952).
(33) Goetzl, F. R., Ahokas, A. J., and Goldschmidt, Margaret, *J. Applied Physiol.*, **4**, 30–6 (July 1951).
(34) Grab, E. G., Jr., and Haynes, D., *Quick Frozen Foods*, **10**, 71 (1948).
(35) Greenwood, D. A., Lewis, W. L., Urbain, W. M., and Jensen, L. B., *Food Research*, **5**, 625 (1940).
(36) Guadagni, D. G., *Food Technol.*, **3**, 404 (1949).
(37) Harris, W. A., Norman, L. W., Turner, J. H., Haney, H. F., and Cotton, R. H., *Ind. Eng. Chem.*, **44**, 2414 (1952).
(38) Hayes, N. V., Cotton, R. H., and Roy, W. R., *Proc. Am. Soc. Hort. Sci.*, **47**, 123 (1946).
(39) Higbie, H., paper presented before Sugar Ind. Engineers, May 1952.
(40) Higbie, H., and Stegeman, G., *J. Am. Chem. Soc.*, **72**, 3799 (1950).
(41) Hinton, C. L., *Intern. Sugar J.*, **51**, 222 (1949).
(42) Hirota, K., and Miyashita, I., *J. Chem. Phys.*, **18**, 561 (1950).
(43) Hockett, R. C., *J. Calif. State Dental Assoc.*, **26**, 71 (1950).

(44) Hodge, J. E., and Rist, C. E., *J. Am. Chem. Soc.*, **75**, 316 (1953).
(45) Hohl, L. A., and Swanburg, Joyce, *Food Packer*, **27** (March 1946).
(46) Hohl, L. A., Swanburg, J., David, J., and Ramsey, R., *Food Research*, **12**, 484 (1947).
(47) Hosoi, K., *J. Agr. Chem. Soc. Japan*, **20**, 9–14 (1944).
(48) Hungerford, E. H., and Nees, A. R., *Proc. Am. Soc. Sugar Beet Technol.*, **3**, 499 (1942).
(49) Joslyn, M. A., *Food Technol.*, **3**, 8–14 (1949).
(50) Joslyn, M. A., and Miller, J., *Food Research*, **14**, 325–39 (1949).
(51) Joslyn, M. A., and Supplee, H., *Ibid.*, **14**, 209–15 (1949).
(52) *Ibid.*, pp. 216–20.
(53) Junk, W. R., Nelson, O. M., and Sherrill, M. H., *Food Technol.*, **1**, 506–18 (1947).
(54) Kertesz, Z. I., "The Pectic Substances," p. 194, New York, Interscience Publishing Co., 1951.
(55) Koch, R. B., Geddes, W. F., and Smith, F., *Cereal Chem.*, **28**, 424 (1951).
(56) Kozin, N. I., and Bessonov, S. M., *Voprosy Pitaniya*, **10**, No. 5–6, 24–9 (1941).
(57) Lea, C. H., "Food Science," p. 221, London, Cambridge University Press, 1952.
(58) *Ibid.*, pp. 228–39.
(59) *Ibid.*, p. 319.
(60) Lecat, Pierre, *Compt. rend.*, **224**, 751–2 (1947).
(61) Lewis, V. M., Esselen, W. B., Jr., and Fellers, C. R., *Ind. Eng. Chem.*, **41**, 2587–91 (1949).
(62) Lockhart, E. E., and Stanford, I. E., *Food Research*, **17**, 404–8 (1952).
(63) Lowe, Belle, "Experimental Cookery," 3rd ed., New York, John Wiley & Sons, 1943.
(64) MacArthur, C. G., *J. Phys. Chem.*, **20**, 495–502 (1916).
(65) McGinnis, R. A., "Beet Sugar Technology," New York, Reinhold Publishing Corp., 1951.
(66) Mack, P. B., *Sugar Research Foundation (N. Y.), Rept.*, 28 (1949).
(67) Malmberg, C. G., and Maryott, A. A., *J. Research Natl. Bur. Standards*, **45**, 299–303 (1950).
(68) Margolina, Y. L., and Voyutski, S. S., *Zavodskaya Lab.*, **14**, 321 (1948).
(69) Mashevitskaya, S. G., and Plevaka, E. A., *Zhur. Priklad. Chim.*, **11**, 511 (1938).
(70) Maudru, J. E., and Paxson, T. E., *Proc. Am. Soc. Sugar Beet Technol.*, **6**, 538–40 (1950).
(71) Meeker, E. W., *Food Technol.*, **4**, 361, 365 (September 1950).
(72) Miller, J., and Joslyn, M. A., *Food Research*, **14**, 340–53 (1949).
(73) *Ibid.*, pp. 354–63.
(74) Miller, W. T., *Food Packer*, **27**, 50, 54 (1946).
(75) Müller, C., *Z. physik. Chem.*, **81**, 483–503 (1912).
(76) Murdock, D. I., Troy, V. S., and Folinazzo, J. F., *Food Research*, **18**, 85–9 (1953).
(77) Nakamura, S., and Hemmi, K., *J. Agr. Chem. Soc. Japan*, **19**, 603 (1943).
(78) Nelson, T. J., *Food Technol.*, **3**, 347–51 (1949).
(79) Newth, F. H., *Advances in Carbohydrate Chem.*, **6**, 83 (1951).
(80) Nitzsche, Max, *Deut. Zuckerind.*, **65**, 431–6, 438–48 (1940).
(81) Noyes, H. A., U. S. Patent 2,443,866 (1948).
(82) Ogston, A. G., *Proc. Roy. Soc. (London)*, **A196**, 272–85 (1949).
(83) Othmer, D. F., and Silvis, S. J., *Sugar*, **43**, No. 5, 32–3 (1948).
(84) *Ibid.*, **43**, No. 7, 28–9 (1948).
(85) Patton, Stuart, *J. Dairy Sci.*, **33**, 904–10 (1950).
(86) Pecego, R., *Bol. leite e seus deriv. (Rio de Janeiro)*, **5**, No. 56, 13 (1952).
(87) Poe, C. F., and Cason, J. H., *Food Technol.* **5**, 490–2 (1951).
(88) Powers, H. E. C., *Intern. Sugar J.*, **50**, 149–50 (1948).
(89) Rauch, G. H., *Food Manuf.*, **17**, 34 (1942); *Nutrition Abstr. & Rev.*, **12**, 68 (1942).
(90) Ray, R. L., Reid, M., and Pearce, J. A., *Can. J. Research*, **25F**, 160–72 (1947).
(91) Richardson, J. E., and Mayfield, H. L., *Montana Agr. Expt. Sta., Tech. Bull.* **423** (1944).
(92) Riedel, L., *Chem.-Ing.-Tech.*, **21**, 340–1 (1949).
(93) Riester, D. W., Braun, O. G., and Pearce, W. E., *Food Inds.*, **17**, 742–4, 850 (1945).
(94) Robinson, R. A., and Stokes, R. H., *Trans. Faraday Soc.*, **45**, 612 (1949).
(95) Schroeder, A. L., and Cotton, R. H., *Ind. Eng. Chem.*, **40**, 803–7 (1948).

(96) Schulte, K. E., and Schillinger, A., *Z. Lebensm.-Untersuch. u.-Forsch.*, **94**, 166–82 (1952).
(97) Shamrai, E. F., *Voprosy Pitaniya*, **10**, No. 3–4, 42–8 (1941).
(98) Shearon, W. H., Jr., *Chem. Eng. News*, **30**, 44, 4606–10 (1952).
(99) Sherrill, M., *Sugar*, **44**, No. 6, 48 (1949).
(100) Shumaker, J. B., and Buchanan, J. H., *Iowa State Coll. J. Sci.*, **6**, 367 (1932).
(101) Sills, V. E., *Ann. Rept. Agr. and Hort. Research Sta., Long Ashton, Bristol*, **1939**, 127–38.
(102) Sowden, J. C., and Schaffer, Robert, *J. Am. Chem. Soc.*, **74**, 499 (1952).
(103) Spencer, G. L., and Meade, G. P., "Cane Sugar Handbook," pp. 188, 215, 8th ed., New York, John Wiley & Sons, 1945.
(104) Spengler, O., Böttger, S., and Seeliger, B., *Z. Wirtschaftsgruppe Zuckerind.*, **86**, 193 (1936).
(105) Steiner, E. H., "Food Science," pp. 292–304, London, Cambridge University Press, 1952.
(106) Stevens, Raymond, *Food Technol.*, **7**, No. 2, 86 (1953).
(107) *Sugar*, **47**, 40 (1952).
(108) *Ibid.*, p. 51.
(109) Sumiki, Yusuke, Yamanaka, Shigeru, Oka, Keijiro, and Takeishi, Atsushi, *J. Agr. Chem. Soc. Japan*, **20**, 89–93 (1944).
(110) Swaine, R. L., *Food Technol.*, **5**, 291–2 (1951).
(111) Taufel, K., and Burmeister, H., *Z. anal. Chem.*, **129**, 352–65 (1949).
(112) Taylor, Millicent, *J. Chem. Soc.*, **1947**, 1678–83.
(113) Tracy, P. H., Sheuring, J. J., and Dorsey, M. J., *J. Dairy Sci.*, **30**, 129–36 (1947).
(114) Underwood, J. C., and Rockland, L. B., *Food Research*, **18**, 17–29 (1953).
(115) Uvarova, A. P., *Sakharnaya Prom.*, **19**, No. 9, 21–3 (1946).
(116) Van Blaricom, L. O., *Proc. Am. Soc. Hort. Sci.*, **50**, 229–30 (1947).
(117) Van Hook, A., *Ind. Eng. Chem.*, **38**, 50–3 (1946).
(118) Van Hook, A., *Proc. Am. Soc. Sugar Beet Technol.*, **6**, 570–5 (1950).
(119) Van Hook, A., and Russell, H. D., *J. Am. Chem. Soc.*, **67**, 370–2 (1945).
(120) Vavruch, Ivan, *Listy Cukrovar.*, **64**, 245–8 (1948).
(121) Weiss, F. J., *Baker's Digest*, **21**, No. 6, 116–20, 123 (1947).
(122) *Ibid.*, **22**, 4–7, 27–30 (1948).
(123) *Western Canner and Packer*, **44** (1952).
(124) Wilcox, E. B., Greenwood, D. A., and Galloway, L. S., *Farm & Home Sci. (Utah Agr. Expt. Sta.)*, **13**, No. 2, 32–3 (June 1952).
(125) Winslow, E. T., "Beet Sugar Technology," McGinnis, ed., p. 391, New York, Reinhold Publishing Corp., 1951.
(126) Wolfrom, M. L., *et al.*, *J. Am. Chem. Soc.*, **71**, 3518 (1949).
(127) Wray, R. R., personal communication, Feb. 24, 1953.
(128) Yamazaki, J., Tsuzuki, Y., and Kagami, K., *Kagaku*, **17**, 175 (1947).
(129) Young, F. E., and Jones, F. T., *J. Phys. & Colloid Chem.*, **53**, 1334–50 (1949).
(130) Young, F. E., Jones, F. T., and Black, D. R., *J. Am. Chem. Soc.*, **74**, 5798–9 (1952).
(131) Young, F. E., Jones, F. T., and Lewis, H. J., *Food Research*, **16**, 20–9 (1951).
(132) Zerban, F. W., Sugar Research Foundation, *Technol. Rept. Ser.*, No. **2** (1947).
(133) Zerban, F. W., and Martin, James, *J. Assoc. Offic. Agr. Chemists*, **27**, 295–302 (1944).
(134) Zerban, F. W., and Martin, James, "Methods of Analysis of Association of Official Agricultural Chemists," p. 812, 7th ed.

RECEIVED April 15, 1953.

Starches in the Food Industry

THOMAS JOHN SCHOCH and ALBERT L. ELDER

George M. Moffett Research Laboratories, Corn Products Refining Co., Argo, Ill.

Fundamental concepts of starch chemistry provide an interpretation of its function and behavior in various food uses. The characteristics of modified starches depend on granule structure and on specific size and shape of the component molecules. Most native starches contain both linear and branched polysaccharides. The linear fraction is responsible for gel formation and for various retrogradation effects, the branched fraction for high colloidal stability and good suspending qualities. Starches are employed in the food industry as gel formers, thickening agents, and colloidal emulsifiers. Desired characteristics can frequently be enhanced by choice of an appropriate modified starch. The various food uses are individually discussed from the standpoint of molecular behavior and optimal types of modification.

Starch represents the chief source of dietary carbohydrate, whether in the refined form of a packaged instant dessert or as the primitive parched corn carried by an Iroquois warrior. With the rise in world population, it becomes increasingly important to employ carbohydrates to the maximum extent as a source of energy, thus conserving the supply of protein. Hence a challenge is presented to make this starchy intake as palatable and as nutritionally efficient as possible, and likewise to adapt its physical characteristics to new and special food uses. Starch in the human diet comes largely from such basic foods as roots, tubers, and the cereal grains. Even in the United States, the amount of refined starch is of minor importance in the diet. Nevertheless, the most important single use of manufactured starch is as a food, far outweighing any of its diverse industrial applications. Indeed, more than 30% of the total starch production of the wet-milling industry is absorbed in food channels, and this figure does not include the large amounts of starch used in the manufacture of sugar and sirup.

Our grandparents knew starch chiefly in the form of cooked cornstarch puddings, which often had a somewhat tough and rubbery texture. During the past few decades, the starch industry has made remarkable progress in improving the quality, character, and acceptability of its food starches. In the earlier stages, progress was made chiefly through a laboriously acquired practical knowledge of how to modify starch to enhance various desirable properties. More recently, an understanding of the fundamental chemistry and physical behavior of starch molecules and granules has begun. This knowledge has materially aided the search for more useful starches in the food industry. It seems advisable to review certain pertinent phases of fundamental starch chemistry before correlating this information with specific food uses.

Physical and Chemical Characteristics

In one form or another, starch constitutes the reserve food supply of all plant life. As stored in the grain of corn or wheat, it provides sustenance to the seed through the processes of germination and growth, until such time as the young plant is capable of synthesizing its own starch. In the tapioca root or potato tuber, it provides a reserve food depot for the plant during periods of dormancy or growth. Microscopic examination of these storage depots shows that the starch is organized into tiny spherules or granules, the size and shape of which are specific for each variety of starch. The general appearance of the six most common food starches is shown in Figure 1. Potato starch has relatively large oval granules, 15 to 100

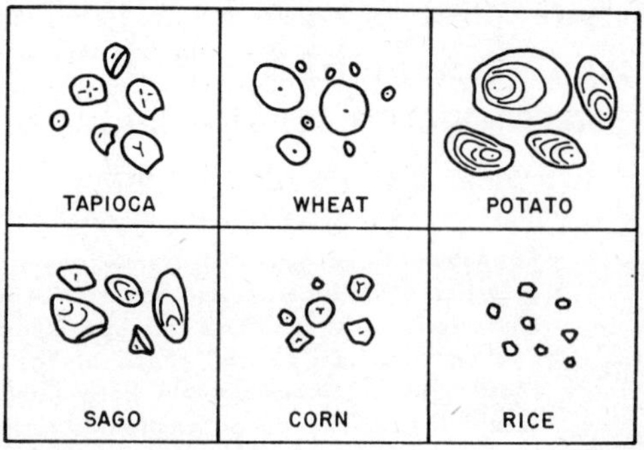

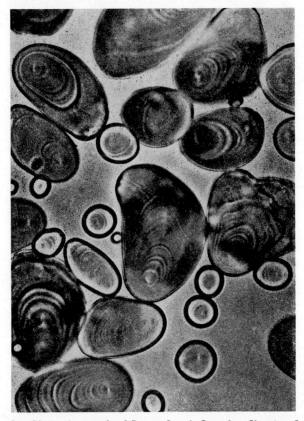

TAPIOCA	WHEAT	POTATO
SAGO	CORN	RICE

Figure 1. Microscopic Appearance of Various Starch Granules

microns in diameter, with pronounced oyster shell–like striations around an eccentrically placed botanical center or hilum. Rice starch is the smallest of the

Figure 2. Photomicrograph of Potato Starch Granules, Showing Striations

common starches, 5 to 6 microns in diameter and polygonal in shape. Cornstarch is likewise polygonal, the result of compacting pressures within the kernel during drying of the corn in the field.

The granule appears to have successive layers or growth rings like an onion and these are particularly well defined in potato starch (Figure 2). When viewed under the polarizing microscope, the granule shows a strong interference cross centering through the hilum (Figure 3). In addition to this concentric organization, the granule appears to have some sort of radial structure, since it fractures into radial segments when crushed between a microscope slide and cover glass. Hence the granule may be considered as a typical spherocrystal. Indeed, the x-ray diffraction pattern of granular starch indicates that crystalline organization extends all the way down to the molecular level.

Granular starch is completely insoluble in cold water. If an aqueous suspension is heated, nothing much occurs until a certain critical temperature is reached.

COURTESY, S. A. WATSON

Figure 3.   Potato Starch Granules Under Polarized Light

Then a whole series of physical changes is initiated. If heated on the hot stage of a microscope, the granules suddenly start to swell, simultaneously losing their polarization crosses. This phenomenon is called gelatinization. Many of the granules show delayed swelling, so that gelatinization actually occurs over a temperature range rather than at a single point. For example, this range is 56° to 67° C. for potato starch, 64° to 72° C. for corn, and 69° to 75° C. for grain sorghum. After their initial rapid swelling, the individual granules continue to expand more slowly as the temperature is raised, until their outlines become indistinct under the microscope. During this process, the granules become so large that they begin to jostle one another, thus imparting the viscous consistency typical of a cooked starch paste. This does not represent actual solution of the starch granules but merely their progressive swelling, as can be demonstrated by use of the phase microscope. These swollen granules behave as individual gel particles which are remarkably coherent and elastic. If the paste is vigorously stirred, many of the swollen granules will be disrupted into fragments, with a consequent decrease in viscosity. But these fragments still persist as intact entities. While a small portion of the total starch substance does go into actual solution during gelatinization and swelling of the granule, the major part persists as swollen granules or fragments thereof. These can be dissolved only by elevating the temperature above 100° C.—e.g., autoclaving a starch paste at 20 pounds' pressure for an hour or two.

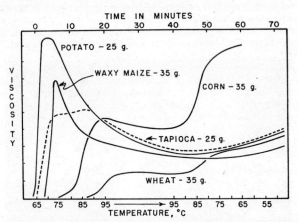

Figure 4.   Brabender Pasting Curves of Various Starches
Paste concentration given in grams of starch per 450 ml. of water

There is a simple and useful way of following the gelatinization of starch, by graphing the changes in viscosity when a stirred suspension of starch in water is heated at a uniformly rising temperature rate. This is conveniently accomplished by means of the Brabender Amylograph or the Corn Industries viscometer *(3, 13)*, which plots increase in viscosity over the gelatinization and swelling range. The pasting curves of various typical starches are illustrated in Figure 4. Tapioca starch frequently shows a peculiar and unexplained double hump during pasting, then thins down markedly as the swollen granules are distorted by cooking and by mechanical agitation. In comparison, cornstarch gelatinizes at a substantially higher temperature, does not give the high peak viscosity of tapioca, but likewise does not show the drastic thinning on continued cooking.

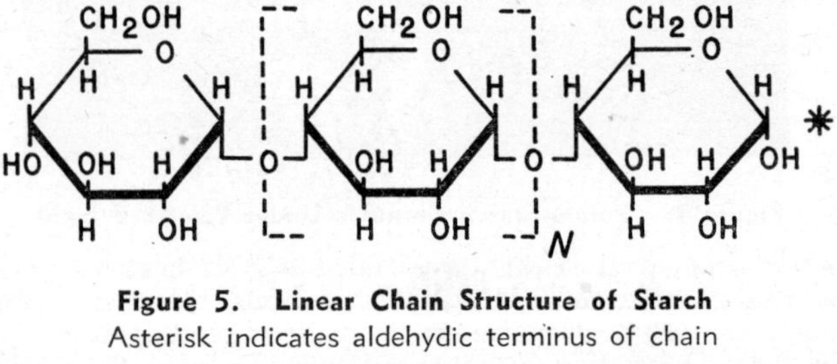

Figure 5.   Linear Chain Structure of Starch
Asterisk indicates aldehydic terminus of chain

So the process of granule swelling represents a tendency of the starch substance to hydrate itself with more and more water, but never quite giving a true solution except by such means as autoclaving. There is an important reverse reaction, whereby starch pastes and solutions show evidence of dehydration and insolubilization. This is manifested in a variety of ways. A bottle of starch indicator—almost transparent when originally prepared—gradually becomes more and more opaque on standing and finally deposits a white sediment. Usually, this sediment cannot be redissolved by heating. If the starch indicator is frozen and thawed, the starch substance is completely insolubilized. If a cooked starch paste is allowed to cool undisturbed in an open beaker, it will form an insoluble skin which cannot be redissolved by heating. Still another example is the spontaneous thickening of a starch paste when it is allowed to stand, in many instances setting up to a rigid irreversible gel, exemplified by the old-fashioned cornstarch pudding. All of these related phenomena are categorically designated as retrogradation, representing the spontaneous transformation of the starch substance to a less soluble or less hydrated state. However, the different species of starch show wide variations in their tendency to retrograde. The interpretation of these phenomena has required extensive investigation into the physical and chemical structure of starch.

Starch is a high polymer, formed in the plant by the progressive condensation of glucose units through the agency of a specific enzyme. In effect, this builds up a long chain containing some hundreds of glucose units (Figure 5). However, there is a second enzymatic process in the plant which is capable of inducing branch points in the synthesis. Hence the two enzymes acting jointly build up a large treelike molecule of perhaps several thousand glucose units, in which the average length of the individual branches is some 25 to 30 glucose units (Figure 6). Most of the common starches—corn, wheat, potato, and tapioca—contain both linear and

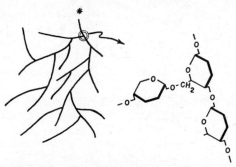

Figure 6.   Branched Structure of Starch
Asterisk indicates aldehydic terminus

branched types of starch molecules. The ratio of these fractions is fairly constant for any given species of starch, 17% of linear material in tapioca starch, 22% in potato, and 28% in corn. But there are likewise the so-called waxy or glutinous cereals, recessive genetic varieties of rice, maize, and sorghum which were first found in the Orient (11). These starches contain only the branched type of starch fraction. In contrast, the starches of wrinkled-seeded garden peas (9, 23) and of certain varieties of sugary corn (6) are composed predominantly of linear fraction.

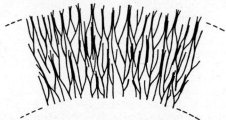

Figure 7.   Micellar Organization Within Starch Granule, According to K. H. Meyer
Micellar regions indicated by thickened sections

High polymeric molecules have habits and behaviors utterly different from the basic monomer units. The influences of molecular size and shape become of primary importance. A solution of a high polymer does not readily splash, because the large extended molecules have too much molecular inertia or viscosity. Large molecules likewise have shape—one-, two-, or three-dimensional—and the particular shape has an enormous influence on physical properties. Many of the characteristics of the individual starch fractions can be traced to considerations of shape, the behavior of flexible linear molecules as contrasted with that of globular bushy molecules.

There is a further specific property of linear high polymers, the tendency of the molecules to cling together and to line up in parallelwise fashion. This effect shows up with the so-called long-fibered greases, where the linear hydrocarbon molecules string out into alignment between the fingers. Kistler *(14)* has described this effect as "orienting a plate of badly confused spaghetti by pulling it from opposite sides." When the molecules are heavily loaded with hydroxyl groups as in starch, the associative bonding between chains becomes very strong, tending to pull these chains into self-alignment by a sort of zipper action. Within certain limitations, the longer the chains, the stronger is the attractive force between them. Hence there is greater associative tendency between the long molecules of the linear fraction than between the relatively short exterior branches of the branched fraction.

Figure 8.   Micellar Organization Within Swollen Starch Granule, According to K. H. Meyer
Micellar regions indicated by thickened sections

The processes of gelatinization and retrogradation can be interpreted in terms of this high polymer association. Meyer *(21)* proposed that the layers of the starch granule are organized in the pattern shown in Figure 7. Both linear and branched fractions are laid down in a radial fashion. Wherever possible, the linear chains and linear segments of the branched molecules associate laterally to give parallelwise bundles which Meyer termed "micelles." A long linear chain may pass through several of these micelles, or the outer fringes of a branched molecule may participate in a number of such micelles. Between these organized areas are regions of looser and more amorphous character, where the chains and branches crisscross in various degrees of randomness. This pattern of locally crystalline and oriented organization provides a satisfying explanation for such phenomena as the polarization cross and x-ray diffraction spectrum of the granule and particularly for its gelatinization in hot water. Below the gelatinization temperature, water merely soaks into the open intermicellar areas. When the gelatinization temperature is reached, the loose associative bonding in these areas is quickly satisfied by hydration and the granule begins to swell tangentially. Swelling continues as the temperature is raised and more water seeps into the intermicellar areas. Some of the short linear chains may be released from their entanglement and leach out into the surrounding substrate, but most of the branched molecules are too deeply enmeshed. Through all this swelling the crystalline micelles remain intact, and so the system eventually becomes an enormously swollen elastic network of molecules, tied together at frequent intervals by the persistence of the micelle (Figure 8). If this paste is subjected to violent shear, as in the Waring Blendor or by high-pressure homogenization, the opposing stresses may actually snap glucosidic bonds, rather than tear the individual micelle apart. This hydrolysis by brute force is a concept which is plausible only in the field of high polymers. The granule owes its existence and its swelling properties to these micellar areas and not specifically to the presence of either fraction, since both waxy starches and high linear wrinkled pea starch form optically anisotropic granules. However, the content of linear fraction does have a substantial influence on gelatinization temperature and swelling behavior of the granule. Waxy maize starch forms a paste some 6° to 7° C. lower than ordinary cornstarch (Figure 4), while wrinkled pea starch gelatinizes at a much higher temperature and does not disperse or dissolve even on autoclaving.

So it would seem that the higher the linear content, the stronger is the intermolecular association within the granule.

This concept of intermolecular association likewise explains the reverse action of retrogradation, zippering together linear molecules to give gels or insoluble precipitates (Figure 9). If the action proceeds slowly as in a 1% starch indicator

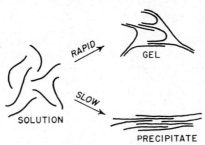

Figure 9.   Mechanisms of Retrogradation of Linear Fraction

solution, the molecules have the opportunity to align themselves in tight combination, and the retrograded material forms an insoluble precipitate. On the other hand, when a hot 5 to 10% cornstarch paste is cooled to room temperature, association is rapid but local and more disorganized, to give an interlacing network or gel structure. In either case, the retrogradation cannot be reversed by such simple means as reheating.

In contrast, the branched starch fraction is much less prone to retrograde. Indeed, in pastes or solutions of the whole starch substance, the presence of the branched fraction has a moderating influence on retrogradation of the linear fraction, slowing down its precipitation and diminishing its gel tendencies. However, the branched fraction does have outer branches some 30 glucose units in length, still capable of some degree of parallelwise association. The principal requirement is high concentration, so that these exterior branches are brought into close juxtaposition. While a 5% paste of waxy maize starch will remain clear and fluid for long periods of time, a 30% paste will harden to a gel on standing. If a 5% waxy starch paste is frozen and thawed, the branched material will be converted to an insoluble state and this product will frequently give an x-ray diffraction pattern indicative of crystalline association of linear chains. The important difference between this sort of association and retrogradation of the linear fraction is the strength of the bonding forces. Retrograded linear material cannot be reversed even by autoclav-

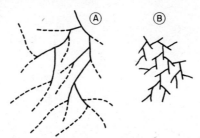

Figure 10.   Branched Starch Fraction (A), Showing Exterior Branches Removed by β-Amylase, and Comparative Structure of Glycogen (B)

ing, while associated branched fraction can be readily restored to its original dissolved state merely by heating to 50° to 60° C. If the branched fraction is treated with β-amylase enzyme, the enzyme acts to remove the outer branches, progressively splitting off maltose until its action is stopped when a point of branching is encountered, as shown in Figure 10. The resulting limit dextrin, with only short stub ends of the original exterior branches, shows no evidence of molecular association. If frozen out of solution, it immediately redissolves when the system is thawed.

Similarly there is the substance glycogen, occurring in animal liver, in certain shell-fish, and in sweet corn. It is the counterpart of the branched starch fraction, except that the branch length consists of only 10 glucose units instead of 25 to 30. In other words, it is a bushlike molecule instead of a tree. Glycogen will not associate or crystallize under any circumstances—its exterior branches are just too short.

The staling of bread appears to be due to crystallization or association of the branched starch fraction in the bread and not to retrogradation of the linear fraction as once thought (28). In support of this theory, canned bread completely staled by 10 months of storage can be regenerated to a completely fresh and edible condition by brief heating to 100° C. Retrograded linear fraction could not be so readily reversed. As further evidence, Geddes and his coworkers (24) have found that bread synthesized from waxy maize starch and undenatured wheat gluten shows a typical staling reaction. The facts seem to fit the theory that the linear fraction is almost completely retrograded during baking and cooling of the loaf, giving the elastic but tender gel structure of fresh bread. Subsequent hardening of the crumb, one of the principal changes which the consumer associates with staling, seems to represent the slow association of the branched fraction.

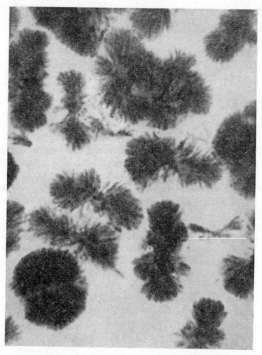

Figure 11. Crystalline Complex of Linear Fraction from Potato Starch
As precipitated with amyl alcohol stained with iodine

The existence of two different polysaccharides in starch was presumed for many years, but largely on indirect and circumstantial evidence. The ultimate separation of pure and undegraded fractions was first accomplished in these laboratories by treating a dilute starch solution with butyl or amyl alcohol (27), which forms an insoluble crystalline complex with the linear fraction (Figure 11). Subsequent work has shown that the familiar blue iodine color of starch is due solely to the linear fraction, which binds iodine as a tight colloidal complex. The branched fraction and the waxy starches merely give a light red coloration with iodine. The branched fraction is stable in solution, subject only to weak association at high concentrations. The linear fractions from various starch species exhibit certain anomalies in behavior which cannot yet be explained—for example, the pure linear fraction from cornstarch shows greatly exaggerated retrogradation tendencies. If a hot 5% solution is cooled to room temperature, it immediately sets up to a rigid, elastic, irreversible gel. In sharp contrast are the linear fractions from potato and tapioca starches, which show only weak gel tendencies. Indeed, the noncongealing qualities of these starches are due primarily to the peculiar character of their linear fractions. It has been variously speculated that their linear molecules are too long to retrograde or that the molecules may be slightly branched. Neither explanation seems entirely satisfying.

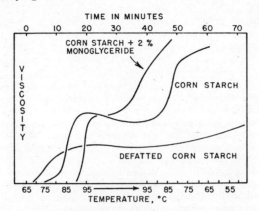

Figure 12. Brabender Pasting Curves, Showing Effects of Defatting
and of Added Monoglyceride

The linear fraction forms insoluble complexes with any polar organic substance of the nature R—OH, R—COOH, R—SO$_3$H, etc. The results of this complex formation are of considerable importance in the food industry, where starch is frequently employed in the presence of monoglycerides or the higher fatty acids. Such materials precipitate the linear fraction when added to a cooked starch paste, resulting in increased opacity, a "short" thick paste consistency, and an almost complete loss of gel strength. When starch is cooked with polar fatty adjuncts, as illustrated in Figure 12, the pasting temperature is raised, swelling is retarded and breakdown and dissolution of the granule are prevented. Commercial cornstarch contains

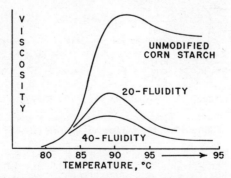

Figure 13. Brabender Pasting Curves of Thick-Boiling Cornstarch and of
20-Fluidity and 40-Fluidity Thin-Boiling Cornstarches

about 0.65% of natural free fatty acid, bound by the linear fraction. If this is removed by alcohol extraction, the defatted starch gives a pasting curve with a lower temperature of gelatinization and a much lower viscosity. The curve can be restored to that of the original cornstarch merely by introducing lipide material into the defatted starch. None of these effects is observed with the waxy starches, which contain only branched fraction. Hence the action of lipide material on pasting of the granule must be attributed entirely to the formation of an insoluble complex with the linear fraction, rendering the granule more resistant toward swelling and hydration.   As an example of this effect, the action of polyoxyethylene monostearate to give a softer bread is attributed to insolubilization of the linear fraction, thereby preventing formation of a gel structure in the fresh bread.

Commercial Modifications

Certain commercial modifications of starch are of interest to the food industry. The simplest type is the acid-modified thin-boiling starch, manufactured by suspending ungelatinized starch in warm dilute acid.   The acid seeps into the more open intermicellar areas, hydrolyzing a few glucosidic bonds here and there.   Superficially, the granule appears unchanged.   However, the intermicellar network has been weakened and when such a starch is gelatinized in hot water, the granule virtually falls apart to give a solution of relatively low viscosity.   Similarly, the action of liquefying enzymes is primarily to weaken the intermicellar structure, thus giving pastes of reduced viscosity.   Most enzymes require that the granule be partially swollen to open up the intermicellar areas for enzyme attack.   These modified starches are used where greater solubility and low viscosity are desired.   The pasting curves of several such starches are shown in Figure 13.   While the molecules of the linear fraction are broken down into shorter lengths, the total content of linear material is not reduced.   Hence these starches cook thin and set up to a gel on cooling, a situation admirably suited to such confections as gum drops.

In contrast to the thin-boiling starches are certain superthick products prepared by chemically cross-bonding the intermicellar areas with very small amounts of such agents as epichlorohydrin or phosphorus oxychloride.   These give ether or ester bridges which supplement the normal intermicellar network.   Consequently, the granule swells more slowly and resists mechanical disintegration, giving a paste of high and stable viscosity.   However, if too many such cross linkages are introduced, the granule does not swell at all.

Heating dry starch with a trace of acid catalyst yields products known as dextrins.   In the earlier stages of dextrinization, the starch is hydrolyzed down to small fragments (Figure 14), which then appear to recombine to give a branched type of structure.   Thus the linear character of the starch is destroyed and the resulting product is highly soluble and stable.   The solubilization of starch which occurs in the crust of bread may represent such a reaction.   In general, these products are of much more interest for industrial purposes than for food uses.

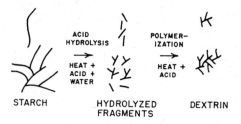

STARCH   HYDROLYZED   DEXTRIN
                    FRAGMENTS

Figure 14.   Mechanism of Dextrinization

In many of the industrial and food uses of the cereal starches, the presence of the linear fraction is undesirable, as it is responsible for gelation, skinning, and excessive thickening.   Such effects can be avoided by using tapioca or waxy maize starch, but these are substantially higher in price and limited in supply.   One alternative is to modify the ordinary starches to reduce the influence of the linear fraction, as by oxidizing the starch with peroxide or alkaline hypochlorite.   While both fractions are affected, the technological benefit is primarily from oxidation of the linear fraction.   Seemingly, if a linear chain is oxidized at two or three points along its length, it becomes warped or distorted by these zigzag discontinuities, so that it will no longer associate or retrograde.   Such modified starches give relatively clear and stable solutions and provide excellent protective colloid action as dispersants and as emulsifiers. Oxidation also causes some sort of obscure scission of the starch molecules, to give products of reduced viscosity. Much the same reduction of linear character can be achieved by slight etherification of the starch, without concurrent loss of viscosity. For example, an alkaline starch suspension may be treated with ethylene oxide to introduce hydroxyethyl groups into the starch. While both fractions are affected, here again the major benefit is in derivatizing the linear fraction. A very minor proportion of derivative groups, perhaps one on every 10th or 20th glucose unit, is sufficient to produce obstructions on the linear chain which interfere with side-by-side association. These starches therefore yield relatively clear and stable pastes which show a minimum of skinning and gelation. Moreover, any desired viscosity can be achieved by derivatizing the appropriate thick- or thin-boiling starch. These and similar products have given excellent results in a variety of industrial uses and are now receiving serious scrutiny for possible application to foods.

The various colloidal functions of cooked starch pastes in food use may be grouped into six general categories: as a thickening agent, as a gelling agent, for moisture retention, as a colloidal stabilizer, as a binder, and as a coating agent. The same starch addition may accomplish several different purposes in any one use. For example, it may contribute a smooth creamy consistency to salad dressing and at the same time maintain the oil component in a stable homogeneous emulsion. Or it may lend both gel strength and moisture retention to such confections as Turkish Delight.

Naturally, there are wide variations in the requirements of any specific colloidal function. For example, starch is employed as a thickening agent to provide a smooth creamy texture to the following products: gravy, cream soups, sauces, chop suey, Harvard beets, salad dressings, prepared mustard, cream pie fillings, fruit pie fillings, pork and beans, cream-style corn, and baby foods.

It is obvious that no single starch product can fulfill all the diverse needs in this listing. Starch systems thicken in a variety of ways, the "long" cohesive stringiness of tapioca or waxy maize, the "short" heavy body of corn or wheat starch, and the viscous flow of a solution of a dextrin or an oxidized starch. The type of thickening action desirable in a baby food is certainly not the same as that for a salad dressing. Choice of the precise consistency is perhaps more of an art than a science. The manner in which a cornstarch pudding feels and liquefies in the mouth cannot be defined with a viscometer or a penetrometer. The user must frequently proceed by trial and error, testing various starch species, various modifications within a given species, mixtures of different starches, and different modes of incorporating and cooking the starch. However, certain generalizations can be made. For instance, all twelve of the products mentioned require a smooth creamy texture and any gel formation would be detrimental. The gel tendency of the linear fraction can be reduced by admixture of tapioca or waxy maize starch, by plasticizing with corn sirup where this is permissible, or by insolubilizing the linear fraction through complex formation with monoglyceride. The latter alternative will, of course, give opaque pastes which are not desirable for such uses as fruit pies. Thorough cooking of any cereal starch improves the clarity of the finished product and reduces its tendency to crack and fissure, probably by increasing granule breakdown to give a more homogeneous system. Where starch and sugar are jointly used, more complete swelling and dissolution of granule structure can sometimes be achieved by adding the sugar after thorough cooking of the starch paste. The transparentizing effect of corn sirup is due primarily to the higher refractive index of the medium.

Tapioca and waxy maize starches require less cooking and yield substantially clearer pastes than cornstarch, partly because of their lower content of linear material and partly because of the peculiar nature of their linear fractions. However, the cohesive stringiness of these starches would be undesirable for most of the cited uses. Moreover, these two starches give very high peak viscosities and then thin out drastically on continued cooking. This pasting behavior makes processing somewhat difficult, particularly in the presence of fruit acid which exerts a further thinning action. The paste may be stabilized by the addition of cornstarch or soft wheat starch, whose granules are more resistant toward mechanical disintegration. In addition, the admixture of cereal starch reduces the long stringiness of the tapioca or waxy maize. Another possibility which is being explored is internal cross bonding within the waxy granule by reacting the starch with trace amounts (0.2% or less) of such agents as epichlorohydrin or phosphorus oxychloride. This chemical reinforcement within the granule eliminates the stringy paste character and stabilizes the viscosity. In addition, the high paste clarity of the waxy starch is retained.

Some original investigations may be of interest at this point. A small amount of food-grade commercial cornstarch is added in the canning of cream-style corn to provide a smooth consistency. Actually, the amount of natural starch in the sweet corn far exceeds this added starch, but most of the natural starch is isolated within the kernel and hence does not contribute to thickening the substrate. Some batches of canned corn may occasionally show curdling, with the separation of a watery serum which gives an undesirable nonhomogeneous appearance to the canned corn. To clarify this question, sweet corn has been fractionated into its starch and protein components at various stages of ripening. At the canning stage, the natural sweet cornstarch was found to have the same content of linear fraction as the added cornstarch. Indeed, the commercial starch gave more translucent pastes that the sweet cornstarch. Hence the small amount of added starch cannot contribute any special instability to the system. However, as much as 9% (on dry basis) of a water-soluble protein which was highly sensitive to heat denaturation was found in the green sweet corn. In addition, the coagulated material from a sample of badly curdled cream-style corn assayed over 70% protein. So it appears that this protein is the offender, though no reason can be offered for the variability in curdling exhibited by different batches of sweet corn.

Some degree of gelation is desirable for certain food uses. The requirements of an instant dessert preparation are fulfilled by a pregelatinized cornstarch or corn-tapioca blend, which reconstitutes with milk to give a soft smooth semigel. Starch must accomplish several purposes in various meat loaf and sausage products—emulsifying and blending with the fat, binding the meat particles, and providing a suitable gel structure. In general, all these functions are accomplished with unmodified cornstarch, cooked directly in the complete mixture. In this connection, transparent films have been prepared from the pure linear fraction of cornstarch and such films might have some possibility as a completely edible sausage casing. However, the lack of a commercially feasible method of starch fractionation is still an obstacle to this development.

Gum confections such as Turkish Delight require a soft elastic starch gel, suitably sweetened and flavored. Unmodified thick-boiling cornstarch gives the traditional qualities to this product and likewise provides a high degree of water retention. Possibly a more familiar form of gum confection is exemplified by gum drops and orange slices. Such products must have a high dry-solids content and good water retention, exhibit good clarity, and have a firm but tender gel structure. These qualities are fulfilled by the use of a relatively large proportion of thin-boiling acid-modified cornstarch, which cooks thin and sets up rapidly to a strong gel. The proper tenderness is provided by the incorporation of corn sirup to plasticize the gel.

There are four minor food uses of starch which exemplify some of its other functions. As a simple instance of water retention, 0.25 to 0.5% of pregelatinized cornstarch is added to meringue and marshmallow toppings solely to prevent syneresis of water, with no effect on consistency of the topping. Use of starch as an emulsifier for fats and oils has previously been mentioned. Similarly, it may be added to chocolate milk drink as a protective colloid to prevent sedimentation of cocoa solids. As another use, panned almonds are coated with a concentrated solution of a very thin starch to seal in the nut with an oil-impermeable barrier. Candy jelly centers are similarly treated to provide a base for subsequent sugaring. A mixture of pregelatinized thick-boiling cornstarch and ungelatinized thin-boiling cornstarch is used in the preparation of wafers and ice cream cones. The pregelatinized starch thickens the batter and the thin-boiling starch imparts crispness, tenderness, and a certain degree of mechanical strength to the baked wafer.

Powdered starch has many large scale food uses: as a dusting agent for bakery goods and chewing gum; as a moisture-sorbing and levigating agent for baking powder, condiments, and powdered sugar; as a flour modifier to reduce gluten strength in cookies, pie doughs, and cake flours; and as a molding form for gum confections. Also, very substantial quantities of starch find their way into the brewing industry and the production of malt sirups.

The food industry has shown an ever-increasing interest in starch as an essential component of processed foods. This widening horizon is evident from the scores of patents covering many phases of starch applications, of which a number of typical examples are listed below.

Current attitudes on the use of chemical ingredients in human food preparations make it imperative to master the utmost potential values in standard food components. This is particularly true of such a material as starch, which has a structural function in foods as well as a nutritional role. The expanding concepts of starch molecules and granules have made possible a more effective adaptation to intended end uses. The authors believe this elaboration of purely scientific knowledge has not reached its zenith.

The manufacturers of cornstarch and its derivatives fully appreciate that cornstarch does not possess all the ultimate virtues and that any single function may be performed by more than one starch species. However, because of the basic role of corn in the American economy, cornstarch may be expected to continue as the outstanding native starch. Its long record of continuous availability is an excellent incentive for the food industry to exploit its virtues and to improve its characteristics through research.

Starch Patents

Among patents in the field of gum confections are those issued to Frischmuth and Frobel (8), Salsburg (26), Schopmeyer (30), Baldwin and Hach (2), Hinz et al. (10), and Olsen (25). Patents on pie fillings have been obtained by Buchanan and Lloyd (5), Lloyd (17), and Arengo-Jones (1). Whitmore and Hickey (32) and Welsh and Rawlins (31) have patents on canning. Schopmeyer (29) has a patent on thermophile-free starch for canning and Lolkema (18) has one on dry infant food. A patent on starch in dry milk products has been issued to Ingle (12).

Among patents in the field of gum confections are those issued to Bergquist (4), Meisel (20), and Lebeson (15). Young (33) has one on icing powder while MacMasters and Hilbert (19) have one on edible starch sponge. A coating for meat products is the subject of a patent issued to Lesparre (16). Musher (22) and Engels, Weijlard, and Schenck (7) have patents pertaining to stabilizers.

Literature Cited

(1) Arengo-Jones, R. W. (to Thomas J. Lipton, Ltd.), U. S. Patent 2,451,313 (Oct. 12, 1948).
(2) Baldwin, A. R., Jr., and Hach, W. (to Corn Products Refining Co.), Ibid., 2,484,543 (Oct. 11, 1949).
(3) Bechtel, W. G., and Fischer, E. K., J. Colloid Sci., 4, 265 (1949).
(4) Bergquist, C. (to Corn Products Refining Co.), U. S. Patent 2,193,470 (March 12, 1940).
(5) Buchanan, B. F., and Lloyd, R. L. (to American Maize Products), Ibid., 2,406,585 (Aug. 27, 1946).
(6) Cameron, J. W., Genetics, 32, 459 (1947).
(7) Engels, W. H., Weijlard, J., and Schenck, R. T. (to Merck & Co.), U. S. Patent 2,232,699 (Feb. 25, 1941).
(8) Frischmuth, G., and Frobel, E. (to Deutsche Maizena Gesellschaft), Ibid., 2,257,599 (Sept. 30, 1941).
(9) Hilbert, G. E., and MacMasters, M. M., J. Biol. Chem., 162, 229 (1946).
(10) Hinz, H. C., Jr., Schermerhorn, G. R., Sr., and Dorn, F. (to American Home Foods), U. S. Patent 2,500,179 (March 14, 1950); 2,554,143 (May 22, 1951).
(11) Hixon, R. M., and Brimhall, B., in "Starch and Its Derivatives," J. A. Radley, ed., 3rd ed., Vol. 1, pp. 252–90, London, Chapman and Hall, 1953.
(12) Ingle, J. D. (to Industrial Patents Corp.), U. S. Patent 2,273,469 (Feb. 17, 1942).
(13) Kesler, C. C., and Bechtel, W. G., Anal. Chem., 19, 16 (1947).

(14) Kistler, S. S., in "Advancing Fronts in Chemistry," S. B. Twiss, ed., Vol. 1, pp. 15–23, New York, Reinhold Publishing Corp., 1945.

(15) Lebeson, H., U. S. Patent 2,266,051 (Dec. 16, 1941).

(16) Lesparre, J. N. (to Armour and Co.), *Ibid.*, 2,440,517 (April 27, 1948).

(17) Lloyd, R. L. (to American Maize Products), *Ibid.*, 2,442,658 (June 1, 1948).

(18) Lolkema, J. (to Scholten's Chemische Fabrieken), *Ibid.*, 2,559,022 (July 3, 1951).

(19) MacMasters, M. M., and Hilbert, G. E. (to United States of America), *Ibid.*, 2,442,928 (June 8, 1948).

(20) Meisel, H. (to Corn Products Refining Co.), *Ibid.*, 2,231,476 (Feb. 11, 1941).

(21) Meyer, K. H., "Natural and Synthetic High Polymers," 2nd ed., pp. 456 et seq., New York, Interscience Publishers, 1950.

(22) Musher, S. (to Musher Foundation), U. S. Patents 2,198,197; 2,198,198 (April 23, 1940).

(23) Nielsen, J. P., and Gleason, P. C., *Ind. Eng. Chem., Anal. Ed.*, **17**, 131 (1945).

(24) Noznick, P. P., Merritt, P. P., and Geddes, W. F., *Cereal Chem.*, **23**, 297 (1946).

(25) Olsen, A. G. (to General Foods Corp.), U. S. Patent 2,508,533 (May 23, 1950).

(26) Salsburg, A. A. (to Dryfood, Ltd.), *Ibid.*, 2,314,459 (March 23, 1943).

(27) Schoch, T. J., in "Starch and Its Derivatives," J. A. Radley, ed., 3rd ed., Vol. 1, pp. 123–200, London, Chapman and Hall, 1953.

(28) Schoch, T. J., and French, D., *Cereal Chem.*, **24**, 231 (1947).

(29) Schopmeyer, H. H. (to American Maize Products), U. S. Patent 2,218,221 (Oct. 15, 1940).

(30) *Ibid.*, 2,431,512 (Nov. 25, 1947).

(31) Welsh, J. F., and Rawlins, A. L. (to American Maize Products), *Ibid.*, 2,191,509 (Feb. 27, 1940).

(32) Whitmore, R. A., and Hickey, F. D. (to Food Machinery and Chemical Corp.), *Ibid.*, 2,592,988 (April 15, 1952).

(33) Young, D. J., Jr. (to Corn Products Refining Co.), *Ibid.*, 2,221,563 (Nov. 12, 1940).

RECEIVED June 3, 1953.

Liquid Sugar in the Food Industry

PAUL R. DAVIS and RICHARD N. PRINCE

California and Hawaiian Sugar Refining Corp., Crockett, Calif.

Sugar solutions have long been in use in the food industry, but more recently they have become available as the liquid sugar of commerce. Sucrose and partially inverted types of liquid sugars and refiner's sirups are now available. Various chemical and physical properties of interest to the food industry are discussed, including solids content and its determination, viscosity, color and color development, and the effect of pH, boiling point elevation, and freezing point depression, ash and its significance, hygroscopicity, and sweetness. Adaptation of various properties to several food processing industries is discussed briefly.

Although liquid sugar and sugar sirups are as old as the sugar industry, only in very recent years have these materials assumed major importance to the food industry as products of commerce. However, commercial availability of liquid sugars has offered food processors cheaper, faster, and better methods of handling a bulk commodity, and these processors have been quick in their acceptance of the liquid products. To assist the chemists and technologists working in the food industry this paper reviews and discusses the properties of liquid sugars and sugar sirups.

Primarily, there are three main classifications of liquid sugars with which food chemists will be concerned: (1) Sucrose types are essentially pure sucrose and, like granulated sugar, contain only a few hundredths per cent ash and other impurities. They are practically water-white in color. (2) Inverted types are also practically pure sugar, but in this instance are mixtures of sucrose and invert sugar, the latter being hydrolyzed sucrose. As in the case of sucrose types of liquid sugar, these mixtures contain only fractional parts of ash and organic nonsugar materials. They are water-white to light straw in color. (3) Refiner's sirup types, although technically not liquid sugars, are included for the sake of completeness. In addition to the sugars present, these products contain varying proportions of ash and associated organic nonsugar constituents. They range in color from a light yellow to a very dark brown or almost black. Flavor characteristics are important.

Types of some of the more frequently encountered liquid sugars are presented in Table I.

Table I. Common Types of Liquid Sugars

Type	Solids, %	Lb./Gal.	Lb. Solids/Gal.	Total Sugars on Solids, %
Sucrose	66.5	11.05	7.36	99.98
90% sucrose 10% invert	68.0	11.14	7.60	99.98
50% sucrose 50% invert	77.0	11.57	8.91	99.97
10% sucrose 90% invert	71.0	11.23	7.97	99.94
Refiner's sirup	77.5	11.66	9.04	92.90

The term "liquid sugar" applies equally to a simple solution of sucrose in water or to a complicated solution which may contain organic and inorganic salts as well as a mixture of several sugars, sucrose, glucose, and fructose. Glucose and fructose, being the hydrolysis products of sucrose, usually occur in equal proportions. Combined in this fashion as invert sugar, they may represent a small portion of the total sugar present or they may be the predominant portion of the mixture. Since each of the components present in the solution will exert its own influence, a rather wide variation in properties exists. Oftentimes these can be utilized to advantage by the food chemist.

Properties of Liquid Sugar

The properties of sugar and sugar sirups have been studied at considerable length over the years, so there is a vast literature available relating to the liquid sugars, particularly in so far as sucrose types are concerned. Food chemists and technologists are referred to well-known handbooks *(3, 10, 12)*, which give these data in detail. For mixtures of the sugars, on the other hand, it is frequently necessary to interpret and expand the information found.

Solids Content. SOLUBILITY. As liquid sugars are sugar solutions, knowledge relating to concentration (or solids content) is of general interest to all food processors. Solids content, while standard for any particular kind of sirup, differs as the proportion of the various sugars changes, since the sugar refiner endeavors to obtain the maximum solids possible in each instance. The importance of obtaining maximum solids is obvious when transportation costs are considered; likewise, in many processes the amount of water to be evaporated is lessened.

The governing factor in so far as solids are concerned is solubility. However, because solubility is a function of temperature, choice of the latter is dictated by the storage temperatures which are apt to prevail. The solubility of pure sucrose at normal room temperature of 20° C. is 67.1% *(2)*. This establishes the normal maximum permissible storage and shipping concentration for sucrose-type sirups if crystallization is to be avoided. A slightly lower value is usually employed in practice to provide some additional margin of safety. In mixtures, on the other hand, as the percentage of invert sugar is increased the amount of total solids which can be contained in solution increases without danger of crystallization of the components. This condition prevails until a total solids content of approximately 80% is reached for a solution containing 33% sucrose and 67% glucose-fructose; thereafter the total solids which will remain in solution drop rapidly. This is shown graphically in Figure 1, adapted from the work by Junk, Nelson, and Sherrill *(8)*.

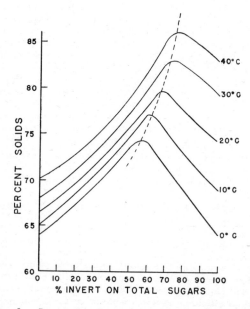

Figure 1.   Permissible Storage Densities for Liquid Sugars

Although these curves showing maximum permissible concentrations are similar at different temperatures, the invert-sucrose mixture for peak solubility varies with temperature. The proportion of invert sugar gradually decreases from near 75% at 40° C. to about 55% at 0° C. Also as the maximum concentration is passed, the effect of temperature on permissible storage density is increasingly marked, but at 50% invert content the effect of temperature is least.

That liquid sugars containing invert sugar can be produced with the higher solids content is due to the fact that fructose is very soluble and that glucose, although least soluble of the three sugars, can exist in solution for long periods of time at saturations well above unity. This is true even though the solid phase of this sugar is also present. For instance, Jackson and Silsbee (7) in studies on honey found average supersaturation coefficients as high as 2.42 with respect to glucose. In this laboratory it has been established that in invert sirups a supersaturation coefficient of 1.75 with respect to glucose is permissible without danger of crystallization, but that the coefficient for sucrose must not greatly exceed 1.0. Figure 1 is based on these coefficients of supersaturation.

With respect to refiner's-type sirups, these follow the same general pattern exhibited by the inverted sirups, since they contain appreciable quantities of the invert sirup; therefore, the total solids found in refiner's-type sirups is of the same order of magnitude as the solids found in the inverted sirups. However, the refiner's-type sirups also contain inorganic salts. These, likewise, often exist in the solution in saturated form and their solubilities differ in the different sugars. This is an important consideration to the sirup blender who changes the relationship of sugars present. Many of the inorganic salts remain in supersaturated state for shorter periods of time than the sugars; this explains the turbidity which sometimes develops in sirups after blending.

Table II. Deviation of Apparent Specific Gravity of Invert Sirups from Sucrose Sirups of Similar Solids Content

Sucrose Sirups		Invert-Sucrose Sirups			
Solids, %	Appar. spec. gravity	Solids, %	Invert on solids, %	Appar. spec. gravity	Difference
74.76	1.3800	74.76	40	1.3761	−0.0039
77.33	1.3970	77.33	50	1.3926	−0.0044
71.24	1.3576	71.24	90	1.3503	−0.0073

DETERMINATION OF SOLIDS. The solids present in liquid sugars can be determined in any of several ways—by drying under vacuum, by specific gravity (either by use of the hydrometer or pycnometer), or by refractometer. These methods have been described in detail (1). Solids by drying is, of course, applicable to all liquid sugars, pure sucrose and mixtures alike. On the other hand, the instruments and tables correlating specific gravity or refractive index to per cent solids are constructed on the basis of pure sucrose solutions. Therefore, they are correct only for the sucrose-type sirups and corrective factors are necessary when dealing with the other type of products.

In this connection, emphasis is placed on the corrective factor which must be applied to refractometer readings when solids in an inverted product are determined by this instrument. This factor is additive, being 0.022 for each per cent invert in the solution, and was first determined by de Whalley (13). The use of this correction has received official acceptance by the Association of Official Agricultural Chemists.

Specific gravity tables refer to pure sucrose solutions. While the other sugars present do not change the specific gravity greatly from the values for sucrose, nevertheless they do change them somewhat. In control work at this refinery the apparent specific gravity of a great number of liquid sugars containing invert has been determined. When compared with the sucrose tables, the solids indicated by these specific gravity tables are somewhat high. This difference becomes increasingly greater as the percentage of invert sugar is increased. Table II shows the observed per cent solids and apparent specific gravity of these inverted products, together with the specific gravity of sucrose solutions of the same solids content. The difference between the two apparent specific gravity values affords a means of approximating the solids in an invert-sucrose mixture from the sucrose specific gravity tables.

Viscosity. Because partially inverted liquid sugars (including the refiner's sirup types) have a greater solids content than sucrose-type sirups, these products will be found to have considerably higher viscosities than the sucrose sirups. The higher viscosities, of course, have considerable bearing on equipment design, particularly storage facilities, pumps, and pipelines. Nelson *(8)* has determined the viscosities of some of the more common liquid sugars. These are shown in Figure 2.

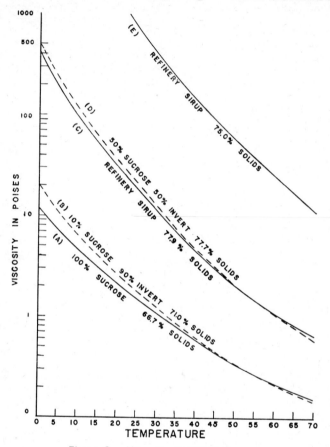

Figure 2. Viscosity of Liquid Sugars

For the less common blends, incomplete viscosity-temperature data can be extrapolated with a fair degree of certainty by the method of Othmer and Conwell *(11)*. Using log-log paper, limited known data are plotted against the corresponding well-established viscosities of a sucrose solution for the same temperature and any convenient density. This gives a straight-line plot which permits ready expansion of the meager data. This method was used to expand the data in Figure 2 from 20° to 0° C.

Table III. Effect of Invert Sugar on Viscosity of Liquid Sugar

Dry Analysis, % Sugar		Viscosity, Cp., at
Invert	*Sucrose*	*22° C. and 65% Solids*
0	100	133
10	90	127
50	50	97

A fallacious impression occasionally has existed that viscosity of a sugar solution can be increased by inversion. This belief apparently stems from the fact that liquid sugars containing invert are prepared at higher densities than uninverted sugar liquors and for that reason are more viscous. However, at the same density, an inverted-type liquid sugar is less viscous than a corresponding sucrose type. This is demonstrated in Table III.

Occasionally, as in the confectionery industry, it may be desired to blend two products of widely differing viscosities, such as corn sirup and liquid sugar. An approximation of the viscosity of such a blend may be obtained by proportioning the logarithms of the viscosities of these products using the antilog as the approximate viscosity of the blend.

As an example, corn sirup with a viscosity of 90,000 cp. was blended with an equal volume of liquid sugar having a viscosity of 235 cp., and also in the proportion of 1 volume of corn sirup to 4 volumes of sucrose sirup. Actual viscosities of these blends are compared in Table IV with calculated viscosities based on arithmetical and logarithmic proportioning. This table shows that the logarithmic calculations for these blends give figures which agree with actual viscosities within a factor of 2, while arithmetical relations give figures which differ from actual values by a factor of 15 to 30.

Table IV. Viscosity of Blended Corn Sirup and Liquid Sugar at 20° C.

	Viscosity, Cp.		
		Calcd. for Blend	
	Actual	*Arithmetical*	*Logarithmic*
Corn sirup	90,000	......	......
50% corn sirup 50% liquid sugar	2,600	45,117	4599
20% corn sirup 80% liquid sugar	620	18,188	773
Liquid sugar	235	......	......

Color Development. The color of a liquid sugar may have very different significance in different food industries. In certain light-colored food products, color corresponding to the lightest granulated sugar may be considered essential. At the other extreme, the darkest colored refiner's-type sirups may be required to augment color. In other instances, color of the liquid sugar may be used primarily as an index of quality. It has long been used as such in the sugar industry. However, in each instance, color may be influenced (and controlled within limits) by changing the hydrogen ion concentration. Figure 3 shows the effect of varying pH value

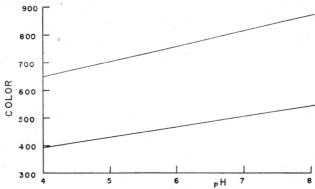

Figure 3. Effect of pH on Color of Liquid Sugar

on apparent color. Two differently colored samples of a 60% sucrose–40% invert blend were selected for this test. An increase of 8 to 10% in color was noted for each unit rise in pH value.

Color development during processing must also be considered. An inverted liquid sugar will develop color more rapidly than a sucrose type. This is illustrated in Figure 4, which compares the color development between a sucrose sirup and one substantially inverted.

Boiling Point Elevation. In contrast to normal saturated solutions wherein the boiling point remains constant and solute crystallizes out of solution as water is removed by evaporation, liquid sugars do not reach a constant boiling point. Instead, the boiling point continues to increase even though the concentration of the sugars passes the point of saturation and the solutions become increasingly supersaturated. The boiling point rise, of course, depends not only on the weight con-

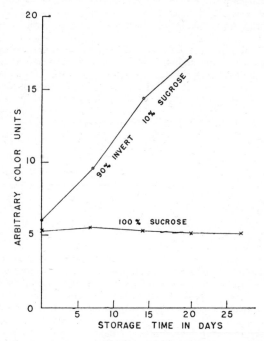

Figure 4. Color Development in Liquid Sugar at 32° C.

centration of the sugars but also on their specific nature. For instance, a molal solution of sucrose has a theoretical rise in boiling point of 0.52° C. (*3*). On the other hand, a solution of the monosaccharides of hydrolyzed sucrose of the same weight concentration would have about twice the boiling point elevation. Mixtures of sucrose and monosaccharides would effect a change intermediate between the two extremes. Boiling point elevations of refiner's-type sirups can also be estimated but because each sirup contains different quantities and varieties of nonsugar constituents as well as the sugars themselves, actual measurement should be made for the specific sirup involved.

Illustrating the value of this property, one needs only to turn to the candy industry for an example. Here temperature is used as a guide to concentration requirements. By cooking to a definitely specified temperature the moisture content of the finished article is controlled with considerable accuracy.

Freezing Point Depression. What has been said with reference to boiling point elevation is true also for freezing point depression. The greater the concentration of sugar in a given solution the greater the amount that the freezing point is depressed. In this instance, also, presence of the monosaccharides as invert sugar exerts a greater depressant effect upon freezing point than does sucrose. Dixon and Mason (*5*), working with dilute solutions, found that the depressing effect of complete hydrolyzation was not quite double the effect of sucrose. Limited experiments in this laboratory indicate that for any given total sugars concentration, the effect is proportional to the quantity of the glucose-fructose mixture present.

Ash. Ash content of sugar products is a property which is used generally as an index of quality. This is particularly useful as a guide, since it can be measured rapidly by conductometric methods. Following the method described by Gillett *(6)*, this determination may be made conveniently at 22.8% solids, corresponding to about 25 grams of solids in 100 ml. of solution.

Ash (or conductivity), however, can serve only as an index and not as an absolute measure of quality. True quality is measured by total nonsugars rather than by only the ionized or inorganic nonsugars present. This may cause considerable discrepancy in comparing sugars from different sources, since the ratio of ash to total nonsugars varies widely between sugars from different geographical locations and from different factories. Also, in liquid sugars, some ash is added in the dissolving medium.

Ash content has somewhat different significance for inverted-type liquid sugars than it does for the uninverted types. Here, by normal practice, acid, such as hydrochloric, is used as the catalyst for inversion, followed by neutralization with sodium carbonate. The net result is a slight increase in the ash content of the final product. This, of course, increases the ash (conductivity) in an amount which has little or no relation to the degree of refining. In any event, the ash content of inverted and uninverted liquid sugars is not comparable.

Hygroscopicity. The moisture-absorbing ability, or hygroscopicity, of invert sugar is very different from that of sucrose or glucose. This is due to the fructose present. Whereas sucrose and glucose absorb little if any moisture until relative humidities above 60 to 65% are reached, Dittmar *(4)* has shown that fructose and invert sugar begin to absorb appreciable amounts of water at very low humidities. He also showed that this is true for small increments of only 2 or 3% invert on total solids, increasing to a maximum at about 20%. Liquid sugars, wherein the quantity of invert sugar is precisely controlled, afford food processors unusual opportunity to regulate moisture absorption through guided addition of this beneficial agent.

Sweetness. Sweetness of liquid sugars is a property which comes in for considerable discussion and one about which there has been considerable disagreement. That this is true is easily understandable. In the first place, sweetness is a subjective test which depends upon personal interpretation. There is no chemical test for this property. In the second place, sweetness probably varies with concentration and it is not unlikely that various sugars in the mixtures affect each other in this regard. At any rate, different investigators have assigned widely differing values to the sweetness of the components of liquid sugars.

Some of the more recent work on the sweetness of sugars occurred during the sugar shortage which the food industry encountered during World War II. The beverage industry in particular was interested and much experimental work was conducted to determine the sweetness value of invert sugar at concentrations found in carbonated beverages. Work by Miller and associates *(9)* indicated that no difference in sweetness could be detected between 10% solutions of sucrose and invert sugar in distilled water or in carbonated beverages made from these sugars. Present consensus seems to indicate that, although inverted liquid sugars have a slightly different honeylike taste than sucrose types, there is no appreciable difference in sweetening power.

Applications in Food Industries

As applied to the food processing industries, the property of liquid sugar which is of uppermost importance is its liquid nature. Being a liquid, storage and handling are greatly facilitated and economic benefits of considerable importance accrue. This same property of being a liquid is also one of the biggest disadvantages of liquid sugar for some uses. Being a solution, liquid sugar obviously cannot be used in dry mixes or in applications wherein the water content exceeds permissible limits. This disadvantage, however, is limited and liquid sugar has found wide application in the canning, carbonated beverage, confectionery, baking, and other industries.

In so far as usage of liquid sugars is concerned, this is best understood by referring again to the types of product available. The sucrose type, of course, finds widest application. Since it differs from granulated sugar only in physical form, it is adaptable to all processes wherein dissolved granulated sugar might be used. The canning industry is, perhaps, the largest user of this type of sirup, for, among the other benefits it possesses, it obviates the need for dissolving dry granulated sugar. Likewise, the carbonated beverage and confectionery fields are large users of this type of sirup.

The inverted type is the next most widely used variety of liquid sugar, utilized for specific advantages which the monosaccharides present. The confectionery field, in which the invert sugar is used as a "doctor" (crystallization inhibitor), and the carbonated beverage industry, in which invert may be used to reduce flavor changes in acid beverages during aging, are examples. The baking industry is another example, inverted sirups being used because of their moisture-absorbing ability.

The refiner's sirup types are somewhat more restricted in usage. Nevertheless, these find application in the ice cream, baking, and confectionery fields where special flavoring requirements are encountered. Refiner's sirups, likewise, are used in quantity in table sirup blending operations.

Literature Cited

(1) Association of Official Agricultural Chemists, "Official Methods of Analysis," 7th ed., 1950.
(2) Browne, C. A., "A Handbook of Sugar Analysis," p. 649, New York, John Wiley & Sons, 1912.
(3) Browne, C. A., and Zerban, F. W., "Physical and Chemical Methods of Sugar Analysis," New York, John Wiley & Sons, 1941.
(4) Dittmar, J. D., *Ind. Eng. Chem.*, **27**, 333 (1935).
(5) Dixon, H. H., and Mason, T. G., *Louisiana Planter*, **64**, 397–8 (1920).
(6) Gillett, T. R., *Anal. Chem.*, **21**, 1081–4 (1949).
(7) Jackson, R. F., and Silsbee, C. G., Natl. Bur. Standards (U. S.), *Sci. Technol. Papers*, **18**, 277 (1924).
(8) Junk, W. R., Nelson, O. M., and Sherrill, M. H., *Food Technol.*, **1**, No. 4, 506–18 (1947).
(9) Miller, W. T., *Food Packer*, **27**, 50–4 (1946).
(10) National Bureau of Standards, *Circ.* **C-440** (1942).
(11) Othmer, D. F., and Conwell, J. W., *Ind. Eng. Chem.*, **37**, 1112–15 (1945).
(12) Spencer, G. L., and Meade, G. P., "Cane Sugar Handbook," 8th ed., New York, John Wiley & Sons, 1945.
(13) Whalley, H. C. S. de, *Intern. Sugar J.*, **37**, 353–5 (1935).

RECEIVED April 15, 1953.

Starch Hydrolyzates in the Food Industry

GEORGE T. PECKHAM, JR.

Clinton Foods, Inc., Clinton, Iowa

Starch hydrolyzates have a number of different properties, some controllable and some not. A wide selection of these products is available to the food processor. It is important that the food processor recognize all the properties characterizing starch hydrolyzates and select the product best suited to his purpose.

The hydrolysis of starch to produce sirups and sugars is big business. Among the various types of products of this class, the acid hydrolyzates are by far the largest group. Data are not readily available on the volume of enzymically converted sirups, but while this is a very sizable quantity, it is believed to be relatively small in comparison with the acid hydrolyzates. The total tonnage of corn sweeteners produced in the United States in 1951 was 1,151,418 tons. This compares with 7,737,000 tons of sucrose refined.

This paper deals with the acid hydrolyzates of cornstarch and sorghum starch. Acid hydrolyzates of other starches are equally adaptable but are of lesser importance in the United States because of economic considerations; the same may be said of malt and maltose sirups made by enzymic hydrolysis.

Kirchoff discovered in 1811 that when starch is boiled in acid, sugar is produced. The process was first used commercially in the United States in 1842, but it did not reach significant proportions until about 1857. Since then it has grown steadily to reach its present sizable volume. Concurrent with the growth of this industry, the science and technology of the process have developed until a large family of well-standardized products, comprising a series differing chiefly in the degree of saccharification, is now available. These products are differentiated by their dextrose equivalent (D.E.), which is the copper reducing power calculated as dextrose and expressed as a percentage of the dry substance.

Starch hydrolyzates fall into two general classes: noncrystallizing sirups, having dextrose equivalents within the range of 30 to 60, and solid sugar products, having dextrose equivalents above 80. The 60- to 80-D.E. range is semicrystallizing, forming two-phase systems, and because of handling difficulties is not commercially produced.

The two major properties used to characterize a particular commercial product are dextrose equivalent and moisture content. In the case of sirups, the moisture content is generally specified in terms of density, expressed as degrees Baumé.

In today's commercial practice, noncrystallizing sirups of 30 to 60 D.E. are available in a range of 41° to 46° Baumé, equivalent to a solids content of about 77 to 88%, and also as dried corn sirup at more than 95% dry substance.

Crude corn sugars of 80 to 92 D.E. are available at dry substance contents of about 82 and 88%. Presumably other concentrations might be produced but these are the regular articles of commerce.

Crystalline dextrose of 99+ D.E. is available as the monohydrate containing an average 8.5% moisture or as the anhydrous product.

Hydrolyzates below about 37 D.E. have a dextrin fraction of such large molecular size that it appears to retrograde slowly out of solution—sometimes in a matter of hours, sometimes in 2 or 3 months. When this occurs, the material has a milky appearance. In some cases this is not objectionable, and hence hydrolyzates are offered commercially at any dextrose equivalent from about 30 to about 60. At about 60 D.E. two new troublesome phenomena appear: An off-flavor and brown color begin to develop, which the industry describes as overconversion; and the dextrose concentration has been increased beyond the solubility limit (for sirups of commercial concentration) so that partial crystallization occurs. In hydrolyzates of 60 to about 80 D.E., crystallization is not sufficient to solidify the whole product to a firm mass. Hence hydrolyzates in this range of dextrose equivalents are not offered commercially, although they could be simply made if the user were willing to accept the color, flavor, and partial crystallization. Above a dextrose equivalent of about 80, at suitable concentration, the sirup will set to a solid concrete and can be chipped or handled as solid blocks. It is not economically feasible to carry the hydrolysis above a dextrose equivalent of about 92. In consequence, the dextrose equivalent of the general commercial range of crude corn sugars is about 80 to 92.

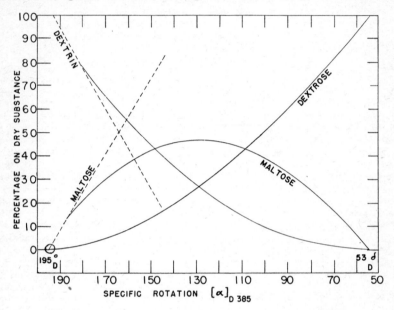

Figure 1. Relation of Carbohydrate Constituents of Acid and Diastase-Hydrolyzed Starch Products to Their Specific Rotation

High dextrose equivalent hydrolyzates can be handled so as to crystallize dextrose and remove the impurities as hydrol. In this way dextrose hydrate is produced, having a dextrose equivalent of 99.6+ and 8.5% water of crystallization. This is a white crystalline compound. If suitable conditions are used, crystalline anhydrous dextrose with substantially no water can also be produced. These two types of crystalline sugar are both regular items of commerce.

Any of several measurable properties can be used for following the course of acid hydrolysis. Two most commonly employed are the copper reducing power and the specific rotation. The chemical composition or carbohydrate distribution in starch hydrolyzates is shown in Figure 1.

Properties of Starch Hydrolyzates

Starch hydrolyzates can be produced of a wide variety of carbohydrate compositions from starch itself to dextrose. All have certain common properties. The intensity of each of these properties is likely to vary with the degree of hydrolysis, but their interrelationships differ considerably. These properties are discussed in five categories: physiological, psychological, chemical, physical, and economic.

Physiological Properties. Physiologically, dietary caloric value is entirely typical of all edible carbohydrates. No nutritive factor other than caloric value is to be expected unless specifically added.

The range of products is all assimilable, the dextrose polymers being readily converted to dextrose by normal body enzymes. Dextrose itself diffuses easily through the walls of the intestinal tract to enter the bloodstream. It is significant that corn sirup has long been a favored ingredient in infant formulas, which attests to its ready assimilability.

None of the products of this class are toxic in the least. Solutions of dextrose are widely used intravenously and corn sirups and even corn sirup dextrins have been so used experimentally with satisfactory results. Diabetic individuals, of course, must use their normal precautions.

Psychological Properties. The products are all palatable, with the exception of the crude sugars which have a bitter flavor, objectionable to most people. The low conversion sirups, around 30 dextrose equivalent, are practically tasteless. As hydrolysis progresses to 35 to 38 D.E., some sweetness is developed. This increases to and through the 60 D.E. product but thereafter is obscured by the increasing bitterness.

When dextrose is crystallized from high dextrose equivalent sirups, the bitter principle stays in the mother liquor and the sweetness of the dextrose itself is unmasked. Dextrose is generally considered about 75% as sweet as sucrose.

Bodying and texturing properties are functions of the type of use and cannot be generalized. Only experience and judgment in a specific application will permit prediction of effect. All products of this class, except the crude sugars having dextrose equivalents of 80 to 92, are compatible with food flavors. They are especially suitable for complementing fruit flavors.

Chemical Properties. From the food standpoint, probably the most important chemical property of the starch hydrolyzates is fermentability. Dextrose, of course, is well known to be readily fermentable by most organisms. Maltose, likewise, is generally readily fermentable. Thus the total of these two sugars roughly defines the fermentability of a hydrolyzate. Physical conditions permitting, the higher the dextrose equivalent, the higher the fermentability.

Sometimes fermentability is sought, as for bread baking and brewing; sometimes it is avoided, as with preserves. Whatever the objective, a suitable combination of dextrose equivalent and concentration can usually be found to do this job.

Probably second in importance among the chemical properties is the ability of the starch hydrolyzate sugars to combine with nitrogenous compounds to produce brown coloration (the Maillard reaction). As in the case of fermentation, sometimes this is sought, as for caramel color manufacture and baking, and sometimes it is avoided, as with candies, icings, and sirups. In general, the higher the dextrose equivalent, the greater the tendency toward browning. This is increased by increasing pH, temperature, time, and the presence of nitrogenous bodies.

Compatibility with other sugars is relatively complete throughout the range of commercial starch hydrolyzates.

Reducing power, so important to the analytical chemist, is not generally significant to food technologists, because it is present only in alkaline media and few foods are alkaline. Nevertheless, reducing power is potentially present and if alkaline conditions develop, it is a formidable property—good or bad.

Acidity, alkalinity, buffer capacity, and pH are all related and are usually readily controllable within limits by the addition of reagents. The pH must generally be kept on the acid side to prevent alkaline degradation with the attendant development of bitterness and brown color. When this reaction takes place, the pH is slowly reduced by the acidic end products until a stable pH is reached. For practical purposes, products are available from pH 3.5 to 5.5 with any reasonable quantity of suitable edible buffer salts, such as acetate, lactate, and citrate.

Physical Properties. The physical properties of starch hydrolyzates are of paramount importance in most food uses.

All of the products are readily water-soluble. Dextrose is the least soluble of all, being soluble to the extent of 55% by weight at room temperature.

While pure dextrose monohydrate has a melting point of 244° to 248° F., the commercial product evidently has some free water since it melts readily in a boiling-water bath (actually at about 180° F.), forming a semistable sirup. Anhydrous dextrose melts at 295° F. In this respect dextrose is different from sucrose, which does not melt, but decomposes at 367° F. This distinction is not generally important in food uses, as water is almost always present and solution occurs.

Corn sirups are noncrystalline glasses in the dry state and as such have no melting points, but instead gradually soften or dissolve in the trace moisture usually present. Strictly anhydrous corn sirups probably could not be produced because more probably intramolecular dehydration would begin before all free moisture had been removed. This, however, is speculative and not based on experimental evidence.

All of the products are somewhat hygroscopic, with increasing hygroscopicity as the dextrose equivalent increases, except that crystalline dextrose, separated from its mother liquor, has relatively less hygroscopicity. Thus the higher conversion sirups and the crude sugars are employed as humectants, whereas crystallized dextrose and dried, low conversion sirups are sometimes used as fillers in dry products.

The dextrin fraction of corn sirup is, in many respects, a vegetable "gum." Where vegetable gums such as tragacanth, acacia, and arabic are employed, corn sirups may well be considered, as they have many of the desirable properties of these more expensive products.

The cohesive power, particularly of corn sirups, is utilized in making pills, lozenges, and the like, and in coatings and icings. This property is also widely used in the adhesive industry.

The optical properties of the starch hydrolyzates are important to the food industry because of the eye-appeal factor, especially in products of very high solids concentration. Dextrose crystallizes in very small crystals and gives an opaque finish. Sirups dry out pretty much to colorless, transparent glasses. Low conversion sirups have a milky appearance from the insolubilization of the higher dextrins. All of the hydrolyzates generally inhibit the crystallization of other sugars, which is sometimes an advantage and sometimes a disadvantage, depending on end use and objectives.

All of the products have a negative heat of solution—that is, they cool a mixture in which they are being dissolved. Conversely, dextrose has a positive heat of crystallization, giving off heat as it crystallizes. The effect is not great and probably not important to most food processors. The negative heat of solution of dextrose is utilized in bakeries for its refrigerating effect in the dough.

The phenomena resulting from molecular weight and count—osmotic pressure, freezing point depression, boiling point elevation, and viscosity—are very important. They follow the expected pattern of behavior. Dextrose, being a monosaccharide of molecular weight 180, has nearly twice the osmotic pressure of sucrose on a weight basis, which enhances its properties as a preservative. Sirup of about 55 D.E. has about the same average molecular weight as sucrose or lactose and hence about the same properties in this area. Lower dextrose equivalent sirups have less effect than sucrose and are frequently preferred in ice cream because they do not depress the freezing point as much as sucrose. Where these properties are important, a careful selection among the hydrolyzates is advisable.

Viscosity is a most important physical property. It also follows the predictable pattern. For a given concentration at a given temperature, viscosity is markedly reduced with increasing dextrose equivalent. At 100° F., a 43° Baumé sirup of 42 D.E. has a viscosity of 14,800 cp., while at the same Baumé a sirup of 55 D.E. has a viscosity of only 9200 cp. These differences are even more accentuated at higher concentrations. At low concentrations the viscosity decreases rapidly and is not generally a factor. Heat, likewise, tremendously decreases the viscosity. The selected data in Table I show generally the extent of viscosity changes.

Table I. Corn Sirup Viscosities

Viscosity, cp.

° F.	42 D.E.	
	42° Baumé	*45° Baumé*
60	122,000	<2,000,000
120	2,750	33,600
180	287	1,730
	55 D.E.	
60	67,200	<1,000,000
120	1,600	16,800
180	167	820

Economic Properties. These do not necessarily vary with either the dextrose equivalent or the moisture but are nevertheless factors which must always be considered in choosing the product to do the job.

Present commercial products sell for lower price on a dry substance basis than cane or beet sugar. Crystalline dextrose is usually priced at a standard differential below refined sucrose. Sirups and crude sugars are priced generally in relation to the corn market but not at a standard ratio; however, they always sell at lower prices (dry basis) than sucrose. Since these two classes of products are priced on different bases, the relationships between them are not constant and thus relative prices sometimes weight the choice among them.

Freight is a most important economic consideration but it is entirely a commercial matter and must be evaluated for each point of delivery.

Cost of handling depends in large degree upon volume of usage. When volume warrants, the most economical method of handling is in bulk—i.e., tank cars or tank trucks for sirups and covered hopper cars for solid products. When lesser volumes are required, the user finds it more economical to buy in packages—barrels or drums for sirups and bags or billets for sugars. Bulk handling equipment is expensive and its cost must be weighed against the potential savings. Suppliers are always willing to consult in this field.

Heat considerations, such as heat of solution, heat of dilution, boiling point or freezing point changes, specific heat, and heat versus viscosity factors, are of a smaller magnitude of importance, but when margins are small they, too, become significant. The corn refining industry is working out these critical data and will furnish them as rapidly as they become available.

Manufacturing Controls

As a regular procedure, the manufacturer controls the dextrose equivalent, which is a measure of the degree of conversion. He also controls the Baumé, which is an empirical function of the density and reasonably defines the dry substance content, although this relationship varies slightly with the dextrose equivalent.

The manufacturer also controls the acidity relationships. As a class, starch hydrolyzates are nonelectrolytes and as such have no acidity, alkalinity, pH, or buffer capacity. In commercial practice, noncarbohydrate materials in the product give these properties and the manufacturer can control these properties by refining techniques or by adding other materials.

Unless otherwise specified, these products are manufactured with minimum color, flavor, and nitrogen content, and maximum clarity.

Special Treatments

Certain modifications of the normal interrelations of properties may be made by special manufacturing techniques.

In some cases secondary conversions with enzymes are applied which reduce the dextrin content and increase the maltose and dextrose. In other cases, hydrolysis may be carried all the way to the desired end point with diastatic enzymes, using either a single enzyme or a mixture or a series.

Refining may be done in whole or in part with ion exchange resins to reduce color, flavor, or mineral content. This improves color and flavor stability but buffer capacity is lost. Occasionally buffers, or even common salt, may be added back.

Heat treatment at high concentration and reduced pH alters the carbohydrate composition, generally reducing the dextrose content if it is over 50% and tending to produce a mixture of dextrose and disaccharides of various glucosidic linkages.

If required, other materials can be blended into starch hydrolyzates to accentuate or diminish specific properties.

To date modifications such as these are rather limited but this is the direction for future research in the starch hydrolyzate field.

Conclusion

The food compounder or processor, when he considers one of these starch hydrolysis products as an ingredient, is looking generally for a single property, be it sweetness, cost, viscosity, fermentability, or any other. But he cannot buy one property. He inexorably gets all the properties. Corn refiners can frequently accentuate or diminish one property but they cannot eliminate the rest. So in the formulation of a process or product it behooves the food processor to remember that all of these properties will be present to some degree. It will be to his advantage to discuss his problem, his objectives, and his desires with a corn refiner who may be able to suggest means to accentuate or diminish a specific property or properties, with a minimum of alteration of the others.

As new or modified food products are developed or as the food compounder becomes more able to define his needs in specific terms, new starch hydrolyzates with optimum properties for the purpose will be developed. There is virtually no limit to the combinations of properties that can be built into starch hydrolyzates and the manufacturers of this class of food ingredient are both able and anxious to cooperate with the user to find and produce the hydrolyzate most nearly perfect for his application.

RECEIVED October 17, 1953.

Pectic Substances in the Food Industries

GLENN H. JOSEPH

Research Department, Sunkist Growers, Inc., Corona, Calif.

A review of the location and function of the pectic substances in plants leads to a better understanding of possibilities for adapting these interesting colloidal substances to food manufacture and of their beneficial role in the diet of man. An effort has been made to clarify the terminology used for the various forms of pectic substances described in scientific literature and trade publications. The most extensive utilization of the pectic materials is dependent upon their ability to form gels, either of the usual high sugar type or of the low solids type which has grown in importance during the past decade. Practical aspects of gel formation with both high and low ester pectinates are discussed, including the mechanism of dispersion or solution in water.

Pectic substances have been components of the human diet since man's first meal of fruits and vegetables in the prehistoric past, so the addition of pectic substances to certain foods cannot be considered as an innovation of the twentieth century. Fruits, berries, vegetables, tubers, twigs, etc., depend upon their content of pectic substances to perform certain of the vital processes which make plant life possible. The food industries have learned only during the past half-century how to compensate for natural deficiencies and variabilities of pectic substances in fruits and berries, and are learning how added pectic materials can improve a wide variety of prepared foods.

A better understanding of this subject is possible when the location and function of the pectic substances in plants are reviewed. The movement of water and plant fluids to the rapidly growing fruits and the retention of form and firmness of fruits are functions of pectin. This intercellular substance in plants is similar in action to the intercellular substance of the vertebrates—collagen (the precursor of gelatin). Protopectin, the water-insoluble precursor of pectin, is abundant in immature fruit tissues. Ripening processes involve hydrolytic changes of protopectin to form pectin and later, as maturity is passed, enzymic demethylation and depolymerization of pectin to form pectates and eventually soluble sugars and acids.

This natural destruction of pectin and consequent loss of turgidity in overripe fruit shows how nature provides for disposal of the rapidly grown fruits. This fact is also of interest to the preserving industry, where it is desirable, for obtaining the optimum flavor and color, to use fully ripe fruits even though the pectin in such fruits has been lowered in quantity and quality by the natural reactions just mentioned. The addition of standardized fruit pectins to fruits deficient in pectin permits jellies and preserves of uniformly good flavor and firmness to be produced.

It is important to note that nature has not only provided enzymic means for the breakdown of pectin in plants but has also provided in the human digestive tract for somewhat similar enzymic action on orally ingested pectin supplied by natural foodstuffs and in prepared foods containing added pectin. Most of the gums are polysaccharide exudates from trees or bushes native to climates where only the rugged and resistant plants can survive. Aldobionic acid, the nucleus of these gums, is very resistant to hydrolysis. Likewise the gel-forming polyuronides from marine plants have a complex nucleus designed to withstand breakdown. This distinction between the ease of digestibility for the pectic substances and the thickening and gelling materials from desert and sea plants is worthy of consideration.

Chemical and Physical Properties of Pectins

The basic unit of the pectic substances is D-galacturonic acid. The structural relationship of this acid to glucuronic and mannuronic acids and the way the galacturonic units are connected in pectin are illustrated in Figure 1. A long

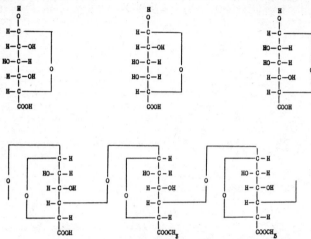

Figure 1.   Structures of Glucuronic, Galacturonic, and Mannuronic Acids

Linkage of galacturonic anhydride units in pectin

chain of perhaps a thousand galacturonic units constitutes what may be termed a pectin "molecule." Such a long chain offers many possibilities for chemical and physical variations. For instance, when all the methoxyl groups are removed from the chain, the substance then becomes pectic acid or a pectate and is of no value as a gel former for the usual type of jelly or preserve. The length of the chains, their thickness, and even the space relationships of the various groups within the chain all have very important effects upon the behavior of the pectins.

The nomenclature of pectic substances, as governed by the extent of esterification, is illustrated by Figure 2. The value of 7.0% methoxyl as the lower limit for

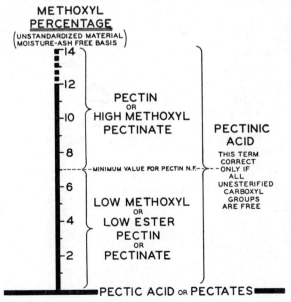

Figure 2.   Nomenclature Chart for the Pectic Substances

what is ordinarily called pectin was chosen by the National Formulary Committee in 1940. It serves as a logical upper limit for the low methoxyl pectins.

The maximum possible methoxyl content for a completely esterified polygalacturonic acid has been shown to be 16.32% (5). Small amounts of natural carbohydratelike materials commonly found associated with pectins, even those made in the laboratory with the greatest of care, tend to reduce this upper theoretical limit to the upper practical limit of 12 to 14%, as indicated by the ordinate of Figure 2. If methoxyl is expressed as a percentage of the polygalacturonic acid present, the presence of natural or added diluting materials in the pectins is thus taken into account. It is only on such a basis of calculation that the 16.32% should be considered as 100% esterification. (Methoxyl and galacturonic acid may be determined easily by methods described under "Pectin" in the National Formulary.) The complex interrelationship between the reagents and the resultant properties of the pectic substances is shown in Figure 3.

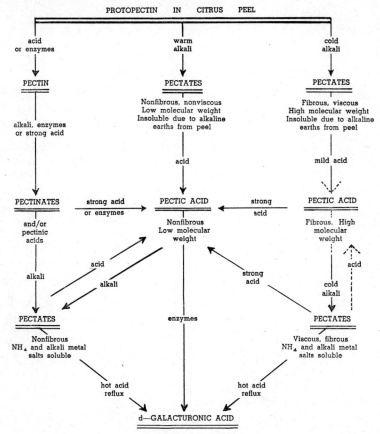

Figure 3. Interrelationships of the Pectic Substances

Commercial pectins are used in foods as thickeners, as gel formers, as emulsifiers, for suspending solids, and for a great variety of specific purposes. All of these uses depend upon the hydrophilic (or water-liking) nature of the pectin and, of course, upon the premise that the pectin is allowed to hydrate and go into solution properly. Considerable financial loss to users of pectin can result from a lack of understanding of the factors which effect the solubility of pectin.

A single and minute pectin particle will go into water solution readily, but when many such particles are added all at once to water, not all can dissolve instantly as single particles. Some of the incompletely hydrated, swollen but undissolved particles adhere to adjacent ones and usually entrap the less completely hydrated particles. The clump thereby formed contains an undissolved core which is protected from the solvent by a slowly dispersed, gummy outer coat. This tendency to clump is easily overcome when some dry, easily soluble substance like sugar is mixed with the pectin to serve as a spacer. Sugar is ideal for such purposes, because it is nearly always an ingredient of food preparations containing pectin. Ordinarily a mixture containing three to five times as much sugar as pectin is used for introducing pectin into water or fruit juices. The pectin can also be made into a slurry with alcohol, glycerol, sugar sirup, or glucose sirup prior to the addition of water, to accomplish the same purpose.

Since solutes (sugar, alcohol, glycerol, salts, etc.) at concentrations exceeding 40 to 50% tend to gel pectin sols, it is evident that if a complete solution of pectin is desired it must be dispersed in aqueous systems of low sugar content; then later, sugar or other soluble solids can be added to the desired level. Attempts have been made by some pectin producers to get around this natural behavior of pectin by incorporating with the pectin small amounts of di- or trivalent metal ions, such as calcium or aluminum. The resulting pectin may be water-insoluble but soluble in slightly acid solutions, such as fruit juices, depending upon the methoxyl and metal ion content of the pectin. So far as is now known, however, this latter type of pectin has never gained much popularity in the food industries.

Gel-Forming Properties of Pectins

Regular Pectin. Ordinary commercial pectins form gels of an entirely different nature and by a different mechanism than gelatin. Aqueous solutions of 1 to 2% gelatin form gels when chilled and pass back into the sol state again merely by heating above the sol-gel transition point, which is below 30° C. An aqueous solution of 0.3 to 0.4% pectin will not gel when chilled, yet when the pH is adjusted to 2.0 to 3.5 and sucrose is present to 60 to 65% of the total weight, a gel forms very quickly, even at temperatures near the boiling point. Gel formation of this type depends upon dehydration and electrical neutralization of the colloidally dispersed and highly hydrated pectin agglomerates or micellar aggregates. Pectin micelles carry a negative charge in ordinary sols. In a practical sense, a dehydrating agent such as sugar, alcohol, or glycerol must be present in amounts of 50% by weight or higher and the pH must be below 3.5. The chains of pectin molecules in gels are considered as being tied together by hydrogen bonding or as being held together electrostatically.

It is important to realize that pectin gels can form at the temperature of boiling sugar solutions, provided the pectin-sugar-acid proportions are correct. The rate of gel formation depends, of course, upon many factors such as sugar and pectin concentrations, pH, type of pectin, and the temperature. When fixed amounts of any one pectin are used, the rate of gel formation becomes more rapid the lower the pH and the temperature and the higher the sugar concentration. Pectins which cause gels to set rapidly are of particular value in the preparation of jams, where it is desirable to have thickening and gelation occur before the fruit rises in the container. Ordinarily, fruit jellies are best when they are not disturbed during the process of gelation. It was for this reason that slow-setting pectins were devised, so that the capping, labeling, casing, and stacking operations could be done before gelation took place.

Commercial preservers should recognize the tremendous effect pH has upon pectin gel formation. When sufficient pectin and sugar are present, no gel will form until the pH is reduced below some critical value near 3.6. This critical pH above which regular pectins will not form a gel, regardless of the amount of the pectin, is sometimes termed the "marginal pH," although more correctly it should be called the limiting pH. The term marginal pH should be reserved for the highest pH at which regular pectins will make a commercially acceptable jelly when used at their designated jelly grades. (The distinction between the words "gel" and "jelly" is important. The word gel is considered as a general term, while jelly applies only to the gels which are commonly known as table jellies and are defined by the Food and Drug Administration as jellies and jams. These latter jellies are always about 65% soluble solids. The soluble solids of gels may be practically zero.)

The curves of Figure 4 show how quickly jelly firmness is lost as the pH goes above 3.3 to 3.4. The four rapid-setting pectins were from four different manufacturers. Many grape juice jellies at pH 2.4 were made from each pectin in order to determine the pectin content required to give jellies, which, after 20 hours of storage at 25° C., exhibited the 23.5% sag on the Exchange Ridgelimeter, considered here as indicating normal commercial firmness. This amount of pectin was then

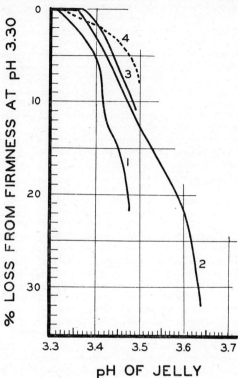

Figure 4. Effect of pH on Jelly Firmness
Grape juice jellies, 1 week old
Four brands of rapid-set pectin

Curve	Pectin
1	Citrus
2	Citrus
3	Apple
4	Apple

held constant for each of the pectins of Figure 4, when subsequent jellies were prepared with decreasing acidity to cover the range of pH practically up to the point of jelly failure. All the subsequent jellies were stored 1 week at 25° C. before they were tested with the Exchange Ridgelimeter.

Curves relating jelly firmness and pH were constructed for each pectin and these curves were then translated to a common point on the ordinate at pH 3.30, to give Figure 4. The increases in firmness during 1 week of standing of juice jellies in this high pH region are also of considerable importance to commercial preservers. The apparent increases in grade attributable to the 1-week standing, for the jellies at pH 3.40 of Figure 4, are shown in Table I.

Table I. Percentage Increase in Apparent Grade of Pectins Attributable to 1-Week Storage of Grape Juice Jellies at pH 3.40

Pectin (Curve No. of Fig. 4)	Source	Increase in Grade, %
2	Citrus	13
3	Apple	17
4	Apple	19
1	Citrus	24

These percentage increases in pectin grade caused by the age of jellies become smaller as the jelly pH is lowered.

Thirty years ago Macara (6) pointed out that jellies at high pH were extremely slow to reach their maximum firmness. The pectin of curve 2 is the best of Table I because it apparently changes less on standing in jellies at high pH than do the other pectins and its limiting pH is remarkably high. A juice jelly at pH 3.45 from the pectin of curve 1 would no doubt be considered a failure when examined 1 day after preparation, yet might be salable a week or two later. An increase of acid to drop the pH only 0.05 would have made a tremendous improvement in the jelly. The curves of Figure 4 illustrate why preservers get great variations in jelly strength, from batch to batch, when they operate in the pH region of 3.3 to 3.5. These variations are sometimes attributed to juice differences but most frequently are blamed on the pectins. Actually, all pectins behave essentially as noted in Figure 4 and the user can obtain more pectin economy and less fluctuation in product firmness by operating in the pH region below 3.3 to 3.4.

Most slow-setting pectins give optimum performance between pH 2.8 and 3.2, while rapid-setting pectins show their best behavior in the region of pH 3.0 to 3.4. Many times an addition of acid to lower the pH only 0.1 unit may be sufficient to prevent complete failure of a jelly batch.

Low Methoxyl Pectin. Four methods have been used to prepare low methoxyl pectins. These methods, using regular high methoxyl pectin as the starting point, partially de-esterify by the use of acids; fixed alkalies or dilute aqueous ammonia systems; enzymes, alone or in combination with alkalies; or ammonia in alcoholic systems or concentrated aqueous ammonia systems. Although each of these methods actually does produce a low ester pectinate, the product resulting from the ammonia method has had acid amide groups substituted for some of the ester groups; thus it differs in character and behavior from the products of the first three methods. As it is the one with which this present paper is concerned, this low ester pectinic acid amide is termed "pectin L.M." hereafter, to avoid uncertainty as to the type meant. The difference in sensitivity to metallic ions and in viscosity of the various pectinates is of sufficient magnitude to prevent one type from being used satisfactorily in recipes designed specifically for another type.

When about half or more of the ester groups of pectin are replaced by carboxyls, as is the case in low methoxyl pectins, the molecule becomes increasingly sensitive to such ions as calcium and ionic bonding or cross linking through primary valences becomes possible.

This ionic bonding gives a gel framework when polyvalent cations are present to tie adjacent chains together. Monovalent cations can, of course, unite with the carboxyl groups and thus prevent bonding to nearby molecular chains. This means that when sodium or potassium is present along with calcium salts, some interference with cross linking takes place and more time for orientation is needed, because calcium cannot then so easily form a random gel structure. Solubility in the presence of calcium may thus be improved, and sometimes when such a salt as sodium citrate is present in low concentrations, a better ultimate gel is the result.

Sugar is not necessary for the formation of gels with low methoxyl pectins nor does sugar interfere with their solubility as is the case for ordinary pectins. High concentrations of sugar, however, tend to interfere with ultimate gel formation with pectin L.M. Ionic bonding gives way to hydrogen bonding as the dehydrating effect of the sugar becomes a factor and the character of the gels is altered. This effect becomes pronounced for sugar contents higher than about 60%.

The acidity of gels with pectin L.M. is not of as great importance as it is with regular pectin gels. Pectin L.M. will make good gels with milk at pH of 6.5 and with fruit or vegetable juices as low as pH 2.5. The desirable range of pH for most salad and dessert gels is 3.2 to 4.0.

The calcium content of low methoxyl gel mixtures is of great importance to users in the food industries. Pectin L.M. is standardized so as to give its normal gel when it is used at 1.0% of the total gel weight and with 25 mg. of calcium ions per gram of pectin L.M. This means that pectin L.M. will not go into solution without boiling when the calcium present exceeds 10 to 15 mg. per gram of pectin. Whenever possible, therefore, the pectin should first be dissolved before any calcium salt is added. When milk is used in making gels with pectin L.M. (custards, puddings, etc.) no calcium salt need be added, since milk supplies a sufficient amount. It is necessary to prepare pilot batches of fruit or vegetable juice gels with varying pectin L.M.–calcium ratios to ascertain the proper balance of these ingredients for the particular product being considered.

The temperature of gelation is of considerable interest when low methoxyl pectin and gelatin are compared, particularly for use in gels for salads and desserts. Gelatin gels form slowly and only when the temperature is below about 28° C. for pure gelatin and about 15° to 20° C. when certain salts are present. This means that good gelatin gels cannot be obtained without chilling in some manner and then the gel will not long endure at summer temperatures. Pectin L.M. gels will form at temperatures above 50° C., even as high as 65° to 75° C. in the case of tomato aspic gels, and the gels will remain satisfactory for table use up to 50° C. and higher.

Standardization of Commercial Pectins

The techniques for standardizing the gel strength of pectins, although far from being entirely perfected, are in a much more advanced state than were those for gelatin after its first 30 years of commercial use. Pectin, like gelatin, is employed in uses so dissimilar and divergent that tests which are adequate for one service are useless for indicating the value in another service. Certain groups of pectin users regard the present schemes of gel strength evaluation as untrustworthy and seem impatient because pectin manufacturers do not agree at once to adopt some of the various use tests which have been proposed. The problem of gelatin grading has been receiving careful study for more than 100 years and is yet not solved in an entirely satisfactory manner. Perhaps pectin grading developments will reach a satisfactory stage in much less time because of the great advances in instrumentation which are now being made. A committee on pectin standardization, appointed by the Institute of Food Technologists, has been working on this problem for several years.

These laboratories, while working on pectin evaluation during a quarter of a century, have realized the futility of trying to assay pectin, even for the food industries, in terms of all its many uses. The goal has been summed up in the following sentences (4):

Jelly firmness was selected as the most necessary and important individual factor for standardization. The goal was to test precisely and reproducibly the jelly firmness which a given pectin is inherently capable of producing under standardized conditions which could be easily maintained in noncritical ranges. Further, the firmness determination should require only rugged, inexpensive equipment and the test must not be time-consuming.

The instrument and standardization techniques which have been devised with these aims in mind have at least permitted uniformity in general jelly-making characteristics to be maintained in a surprisingly constant fashion. Pectin users who once establish recipes and techniques for their own particular products may then rely upon uniformity of behavior for subsequently purchased pectins which have been so standardized. The simple and inexpensive instrument used for measuring jelly firmness is known as the Exchange Ridgelimeter (1) and more than 100 are in use in various parts of the world. Although the present method and instrument are adequate for ensuring uniformity through evaluation of jelly firmness, these laboratories are ever alert to improvements and possible new methods which will benefit both the pectin manufacturer and the ultimate user of the pectin.

Other Pectin Uses in the Food Industries

It is natural that pectin, a normal constituent of fruits and berries, should be used with these foods when thickening and gelation are required. Although pectin does give nearly the same caloric effect upon digestion as does sucrose, it imparts a desirable body and viscosity to fruit juices at a much lower concentration than does sucrose. Knowledge of this fact is used in many canned fruits and berries prepared especially for dietetic purposes.

The thickening of malted milk beverages is easily accomplished by rather low levels of pectin which can be mixed in previously with the malted milk powder. Pectin has been used to suspend solids in cocoa beverages. The emulsifying property of pectin has led to the use of pectin in mayonnaise and salad dressings, in flavoring oil emulsions, and even in cream whipping aids. Pectin, particularly the low methoxyl form, may become widely used in the frozen fruit and berry field during the next few years. Recent publications *(2, 3, 7)* have shown low methoxyl pectin to be desirable for use in frozen strawberries to prevent excessive leakage of juice from the thawed berries.

Although the past 30 years have seen the use of pectic substances in foods grow to the point where several million pounds of pectin are thus used yearly, the future will no doubt bring forth a variety of new pectic substances and a widely expanded field of use in foods.

Literature Cited

(1) Cox, R. E., and Higby, R. H., *Food Inds.*, **16**, 441 (1944).
(2) Grab, E. G., Wegener, J. B., and Baer, B. H., *Food Packer*, **29**, 39–43 (1948).
(3) Joseph G. H., "Pectin L.M.," trade booklet, p. 12, Ontario, Calif., Sunkist Growers, 1947.
(4) Joseph, G. H., and Baier, W. E., *Food Technol.*, **3**, 18–22 (1949).
(5) Joseph, G. H., Kieser, A. H., and Bryant, E. F., *Ibid.*, **3**, 85–90 (1949).
(6) Macara, Thomas, Council on Foods (London), *Research Rept.* **12**, Suppl. 2, 28 (July 1923).
(7) Wegener, J. B., Baer, B. H., and Rodgers, P. D., *Food Technol.*, **5**, 76–8 (1951).

RECEIVED March 17, 1953.

Carbohydrates in Confections

JUSTIN J. ALIKONIS

Paul F. Beich Co., Bloomington, Ill.

The large variety of products turned out by the confectionery industry require a remarkable flexibility in the carbohydrates that serve as major constituents of these products. Physical and chemical properties play a major role. Technological control of raw materials, formulations, processing, and finished confections is important. In choosing the various types of sweeteners the candy technologist is governed by the consumer, for it is only repeat sales that tell confectioners whether they are satisfying consumers.

A bout 70 farm products of varying carbohydrate content furnish over 90% of all the raw materials for over 2000 varieties of confections. This paper is limited to the sweeteners used in confections, which on a dry basis are practically pure carbohydrates. Common sweeteners used in confections are refined crystalline cane or beet sugar, brown sugars, liquid sugars, corn sirup, dextrose, sorbitol, starches, molasses, honey, and maple sugar.

The 2000 varieties of confections seem like a large number, and probably the ACS member has sampled only a fraction of these. If he is an average citizen, he has consumed about 18 pounds of candy annually and this adds up to over 2 billion pounds of carbohydrates to put our industry in a billion-dollar-a-year class. To complicate the problems further, most of the 2000 varieties have hundreds of different variations. There are as many as 500 different formulas for nougats, and probably double that number of formulas for marshmallow. This situation provides a wide opportunity for candy research and development, which is growing rapidly in importance.

This report tries to convert trade terminology into technical terminology, and dwells on the functions of carbohydrates in confections and the reason for the candy technologist's choice. The candy man is charged with producing a type of confection, with the highest quality, at a minimum cost, to have a long shelf life in all types of weather conditions, be attractively wrapped, be consumed by the customer with a satisfaction so as to return for a repeat sale, and, incidentally, be sold at a profit. Candymaking has been an art and will remain an art in many small shops throughout the country.

Considerable credit is due thousands of candy men, who, with no scientific training, have produced confections which are scientifically balanced in formulas and procedures. Many of the secrets of the old candymaker were known physical and chemical phenomena. As in other food fields, there is no magic in candymaking, for there is a reason for everything, but these reasons are often very complex because of the many factors involved (8).

To evaluate effectively the functions of carbohydrates in confections, it is appropriate to consider first the composition of the commercial types and grades of sweeteners and jelling agents (Tables I to IV).

The major factors considered by a candy technologist in determining sweetener usage are differences in physical and chemical properties of various sweeteners, their relative prices, in some cases restrictions imposed by federal or state regulations and, to a lesser extent, advertising and sales programs, in-plant handling problems, consumer preference, and psychological factors.

Table I. Typical Composition of Sucrose Products[a]

	Granulated Sugar		Uninverted Sirups		Inverted Sirups	
	Medium	Fine	No. 1 sirups	No. 2 sirups	Partially	Totally
Moisture, %	0.02	0.04	33	33	27	24
Total solids, %	99.98	99.96	67	67	73	76
Sucrose, %	99.95	99.90	66.6	66.3	42.5	3.5
Invert sugar, %	0.01	0.03	0.1	0.4	30	72
Ash, %	0.005	0.01	0.008	0.08	0.1	0.12
pH			6.5–7.0	6.5–7.0	5	5
Relative	Clear	Clear	Clear	Clear	Water white	Water white
Solution color	Water white	Water white	Light straw	Light straw	Light straw	Light straw

[a] Based upon eastern cane refiners' products.

Table II. Typical Composition of Corn Sirups[a]

		Types of Sirup			
		Regular	Intermediate	High	High
Conversion		Acid	Acid	Acid	Acid enzyme
Baumé		43	43	43	43
Refining		Carbon	Carbon	Carbon	Carbon and ion exchange
Moisture, %		19.7	19.05	18.9	18.5
Color	Lovibond	0.81	0.55	1.3	0.85
Heat color	Series	2.7	2.37	3.0	2.9
D.E.		42.6	51.8	54.6	62.3
pH		5.0	4.98	4.84	4.96
SO_2, p.p.m.		30	26.2	22.5	19.25
Copper, p.p.m.		2	1.4	2.6	.975
Iron, p.p.m.		3	3.22	2.75	1.8
Carbohydrates (as is), %		80.1	80.8	80.8	81.5
Carbohydrates (dry basis), %		99.7	99.7	99.7	99.7
Dextrose (as is), %		17.5	25.9	27.06	29.88
Maltose (as is), %		16.7	20.5	21.88	27.44
Higher sugars (as is), %		16.2	10.95	10.76	12.88
Dextrins (as is), %		29.7	22.7	20.8	11.6
Viscosity (centipoises) at 100° F.		14,000	8970	9090	7350

[a] Based upon average analysis of eight principal corn sirup producers.

Table III. Composition of Refined Corn Sugar (Dextrose)

(Monosaccharide formed by complete hydrolysis of cornstarch)

	Dextrose Hydrate	Anhydrous Dextrose
Moisture, %	9.0	0.25
Dextrose (dry basis), %	99.9	99.9
Ash	Trace	Trace
Form	Crystals	Crystals
Sweetness (basis sucrose 100%), %	70	75
Color	White	White

Table IV. Confectioners' Starches

A. Molding and dusting starches (forms physical shapes of confections). Unmodified cornstarch in powder form, 99% of which will pass 200-mesh sieve
B. Oil-treated molding starch (uses same as A). Powdered starch blended with 0.1 to 0.5 unsaponifiable oil
C. Thin boiling starches (ingredients of candy). Acid modified cornstarch designed to cook out at lower viscosities and set back to firmer gel strength. Available in pearl or powdered forms

The most important physical and chemical properties, considered in their order of relative importance to the candy man, are: (1) relative sweetness; (2) solubility and crystallization characteristics; (3) density of liquid sweeteners and moisture content of solid sweeteners; (4) hygroscopicity; (5) flavor; (6) fermentation and preservative properties; and (7) molecular weight, osmotic pressure, and freezing point depression *(19)*.

Sweeteners vary considerably with respect to these properties, as illustrated in the tables. Requirements also vary widely, according to the qualities desired in the particular type or class of confection. The large variety of products turned out by the confectionery industry require a remarkable flexibility from the carbohydrates which serve as major constituents of these products. There have been some excellent papers on the sweeteners *(11, 13, 15)*.

Categories of Confections

For purpose of this discussion, confections are divided into three categories: hard candies or high boiled sweets, chewy confections, and aerated confections. The second and third categories are further subdivided into two classes, where the sugar solution is supersaturated (grained) or unsaturated (nongrained). Candies that grain, which are of crystalline structure, include the fondant types, such as cream centers, crystallized creams, fudge, pulled grained mints, rigid grained marshmallows, and soft and hard type pan centers. The nongrained candy group consists of marshmallows, taffies, chewy candies, such as nougats, caramels, and molasses kisses, jellies, and gums. There are many intermediate or hybrid types of confections combining the characteristics of both the grained and nongrained candies *(9)*. Sugar is the universal graining agent. The regulators are corn sirup, invert sugar, sorbitol, and others that retard or prevent sucrose crystallization.

Basic Properties of Individual Sweeteners

Sugar (sucrose) has the highest rate of solution, and the smaller the crystal size the more rapid the solution. It forms highly supersaturated solutions which withstand supercooling. Most important of the chemical reactions of sugar is its ability to hydrolyze to produce invert sugar. The color developed in sugar solutions is the function of the pH, and color cannot be formed unless inversion has first taken place. Invert sugar sirups are known for being hygroscopic and retarding the crystallization of concentrated sugar solutions *(16)*.

Corn sirups are noted for retarding and controlling the crystallization of concentrated sugar solutions. This was formerly thought to be due to the dextrins present. Experience has shown that for practical purposes regular corn sirup, high conversion corn sirups, enzyme sirups, and invert sugar have equal effects and can be used interchangeably on a solids basis to control sugar crystallization in such products as fondant, fudge, or other grained confections. Thus dextrins are not necessarily the controlling agent. In physical chemistry protective colloidal action was the explanation for the action of the dextrins. Total solubility of the many sugars found in corn sirup and moisture present in the confection minimize the crystallization of the sucrose. Singly each sugar increases the tendency to crystallization, but this factor of total solubility in the liquid phase overbalances any such tendency. The important thing about corn sirups is that they affect the crystallization of cane sugar to a greater degree than invert sugar and without introducing excessive hygroscopic qualities. A candy man can produce most confections without corn sirup, but the sales hazards would be tremendously increased. Dextrins add viscosity, which increases the body of nongraining confections.

Dextrose possesses the ability to change solubility characteristics and modify the relative sweetness of confections *(12)*. Dextrose also tends to crystallize more slowly than sucrose and the solution at the same concentration is less viscous. Sorbitol, derived from dextrose, is a sugar alcohol, which seems to have plasticizing properties on confections. Besides having a narrow humectant range, it has an effect on the sugar crystal which results in keeping candies soft for extended periods of time. It appears to be gaining in favor as a softening agent for not only grained but nongrained confections *(1)*.

Sweetness

Sweetness is a factor in candymaking and something of a problem. Do customers eat candy because it is sweet, or are some confections a bit too sweet? In turn, is sweetness a detriment to greater consumption? The relative sweetness of maltose, dextrose, and corn sirups, using sucrose as a standard, varies directly with the concentration and is almost twice as great in high concentrations as in low concentrations. A significant supplementary effect of dextrose and corn sirups in combination with sucrose was found (5). This effect is sufficient to give dextrose and high conversion corn sirup the same sweetness as sucrose in 25% solutions consisting of two-thirds sugar and one-third dextrose or corn sirup. Regular corn sirup is as sweet as sucrose in a 40% solution consisting of two-thirds sugar and one-third corn sirup. Thus the relative sweetness of other sugars, compared with sucrose, varies with the concentration.

In a mixture of sugars, if the sweetness values of the components are calculated in terms of dextrose sweetness, the values become additive rather than supplemental as they appear to be when calculated in terms of sucrose sweetness as the standard (3). The sugar and corn products industries, through the nature of their products, have given the confectioner a flexibility and a challenge to his creative abilities for producing new candy.

Hard Candies

Hard candies or high boiled sweets are essentially a highly supersaturated, supercooled solution of sucrose containing 1% or less of moisture. This solution, which may be likened to glass, is usually prevented from crystallizing by the addition of invert (or of invert formed in the batch) or corn sirup, or both, depending upon the properties desired. They are usually compounded in the ratio of 70 sugar to 30 corn sirup for open fire cooking, and 60 sugar to 40 corn sirup in vacuum-processed batches. If the hard candy is to be relatively slow dissolving, a larger proportion of corn sirup is used, as in a banana caramel, which the kids want to last all day; if a quicker dissolving product is wanted, such as a clear mint, smaller amounts of corn sirup are used. Corn sirup controls sweetness and reduces friability of hard candies, which are susceptible to fracture from mechanical shock, because of internal stresses induced by unequal cooling. Sirups of high dextrin content and low dextrose and maltose fractions are used to control hygroscopic properties. Formulations and processing techniques in the manufacture of hard candy confections are designed to produce a product of low hygroscopicity, and different degrees of sweetness, ungrained, dense and brittle in texture.

Chewy Confections

Chewy confections or the nongraining type, such as kisses, caramels, gums, and jellies, are formulated with sucrose, corn sirup, fat, and milk solids. The ratio of sugar solids to corn sirup solids, plus 12 to 15% of moisture, is such that the carbohydrates remain in solution. The dextrins impart the body or chewy texture and in caramels the milk solids contribute flavor and texture. Natural and artificial flavors are accentuated by the sugar, invert sugar, and corn sirup. Fats, which are usually of the vegetable type because of their excellent shelf-life properties, are added to impart body and lubricating qualities. Emulsifiers such as lecithin, monoglycerides, Span 60, and Tween 60 are added to make the product more palatable (17). Standard 42 dextrose equivalent (D.E.) corn sirup has widest application, as it contributes the desirable chewy characteristic without danger of excessive hygroscopicity.

Gums and Jellies. Gums and jellies may be subdivided into two classes: those utilizing starch as the gelling agent, and those utilizing pectin as the gelling agent. Starch gums employ about 10% thin boiling starch which is gelatinized during processing by the free water in a sucrose–corn sirup solution cooked with it. It is essential to use excess water at the beginning of cooking to ensure complete gelatinization—i.e., about 1 gallon of water for every pound of starch in the batch. No excess water is necessary in a new continuous method (4). Sucrose and corn sirup are present in approximately equal proportions and are concentrated to about 76% solids before being cast into a cornstarch molding medium. On cooling, a firm, resilient, transparent gel results. Crystallization is prevented because of the high solubility of mixed sugar present and gelatinized starch. Although standard 42 D.E. corn sirup is in most common usage for gum work, the enzyme-converted 53

D.E. type has advantages due to its humectant properties and extra sweetness. Up to 25% of the total sweetener may also be in the form of refined corn sugar, which promotes sweetness, and reduces viscosity of the batch, enabling faster moisture evaporation and prolonging shelf life because of its hygroscopic character. However, a high ratio of dextrose to sucrose may cause graining in the piece due to the relatively low solubility of the former (10). The introduction of corn sirup has revolutionized the commercial production of gum work.

Pectin and low-methoxyl pectin jellies are compounds of sucrose, corn sirup, pectin, citric acid, water, and a buffer salt, concentrated by heat to a solids content of about 75%, which is necessary for proper gel formation. A 50-50 mixture of sucrose and corn sirup is conventional practice. This ratio effectively inhibits graining and gives the minimum difficulty from excess moisture absorption on the surface. Because the pH of pectin jellies must be adjusted to between 3.45 and 3.55 for proper setting of the pectin gel, significant quantities of invert sugar are formed from the sucrose, giving rise to possible subsequent sweating problems. Accordingly, a corn sirup of relatively low dextrose equivalent is the logical selection, so as to avoid undue amounts of hygroscopic sugars in the mixture. Sorbitol appeared to be effective in all pectin jellies, whereas the MYRJ emulsifiers were more beneficial in starch gum confections (14).

Unsaturated solutions, such as frappés and marshmallows, are not necessarily chewy, although tough marshmallow bars are certainly chewy. Both types of these products are formed with sugar and corn sirup in percentages rarely in excess of 55 to 45, respectively, with water and an aerating agent, usually albumen or soy bean protein for frappés and gelatin for marshmallow work. A new diffusion method for producing foam candy continuously results in longer shelf life because of smaller uniform air cells, and less gelatin or other whipping agents are required because of lessened "fatigue" in process (7). Because of the high proportion of corn sirup and the protection afforded by the protein colloids, the system remains unsaturated with respect to sucrose. Concentration of the total solids without incurring supersaturation to a level safe from microbiological spoilage is made possible with the corn sirup or invert fraction, which may be either standard 42 D.E. or special 52 D.E. type in the case of uncoated goods, or special 52 D.E. and 63 D.E. enzyme or invert sugar in the case of coated marshmallow products. Molasses, honey, or maple sugar is used in all types of confections, mainly as a flavor. Honey has very good humectant properties.

Aerated Confections

Aerated confections are the supersaturated solutions that form grained confections such as creams, fondants, nougats, fudges, and grained marshmallow items. Fondants and creams are prepared by concentrating a sucrose, corn sirup, and water solution to around 85% total solids, using a ratio of sucrose in excess of that of corn sirup so that precipitation will occur when the supersaturated solution is seeded or crystallization induced by mechanical means. This generally implies a proportion of 80 to 70 parts of sucrose to 20 to 30 parts of corn sirup. The mixture is usually boiled to 238° to 242° F. Agitation or seeding is carried on at reduced temperatures of around 110° F., so that the crystals so formed will be of impalpable size. Some air is incorporated by the mechanical action of the equipment used in making the fondant. In the case of creams, the fondant is used as a seeding medium and a small percentage of egg or soya bean albumen in the form of mazetta is used to import some air. The function of the corn sirup portion is to serve as a humectant that will keep the products soft and palatable, and to permit concentration of soluble solids in the liquid phase to a level of around 80%, which will prevent growth of microorganisms. All three types of corn sirups have applications, the special 52 D.E. and enzyme–converted 63 D.E. imparting effective humectant properties. Invert in combination with corn sirup gives excellent results. Refined corn sugar may be used in limited amounts, but it is not generally recommended because of its tendency to grain off in coarser crystals. Sorbitol with corn sirup seems to give a more uniform and whiter cream and fondant.

Fudges are similar to caramels, containing sugar, corn sirup, milk solids, and vegetable fats, but are slightly aerated by mechanical agitation with the aid of egg or soybean protein frappés. Soybean proteins are gaining favor because they bring out the color of the cocoa powder or cocoa liquor used in chocolate fudges. Ratio of sugar to corn sirup is higher, enabling precipitation of sugar crystals to form short, crystallized texture. All three types of corn sirups have applications, as in creams. Although the sirups of higher dextrose equivalents are used in chocolate-coated bars, the quartermaster ration chocolate-type covered disks have 10% sorbitol to ensure a shelf life of the product for at least 2 years *(6)*.

Grained marshmallows are produced with sugar, corn sirup, and an aerating agent, such as gelatin or albumen. Again the corn sirup, invert sugar, or sorbitol serves to retain softness in a supersaturated, aerated sucrose system. Refined corn sugar (dextrose) may be incorporated to 20 to 25% of the total sweetener because of its hygroscopic qualities and higher fluidity qualities, permitting faster beating to the desired specific gravity of the product. Standard 42 D.E. and special 52 D.E. sirups are preferred for this item. Enzyme-converted sirups or invert sugar, if used, should be used sparingly to avoid excessive softening due to absorbed moisture. What holds true for grained marshmallows will hold true for grained nougats. In this confection gelatin is seldom used and the aerating agent is usually egg albumen, soybean protein, or a 50-50 mixture. Vegetable fats are used to promote smoothness and palatability to the confections. Since a high percentage of nougat confections are coated with chocolate or chocolate-type coatings, the 63 D.E. enzyme-converted corn sirups or invert sugar are usually used.

Chocolate coatings are mixtures of sugar, cocoa butter, chocolate liquor, emulsifiers, and flavor, and in some cases milk powder. The function of carbohydrates in confectionery coatings is practically limited to refined cane and beet sugars. A small percentage of dextrose and corn sirup solids is used on some special coatings, but these sweeteners tend to raise the viscosity of the coatings which call for more expensive cocoa butter or vegetable fat. In recent years, because of the commercial grind coatings used on candy bars and on coatings for syndicate outlets in which there is a coarser particle size distribution, as far as the dry ingredients are concerned, more sugar and less vegetable cocoa butter and fats are being used. The sieve-size particles have less surface area than subsieve range particles, thus using less fat and producing a lower viscosity in the coating, which results in considerable savings *(18)*. The intensive use of "chocolate-type" coating which uses cocoa powder as a rule instead of chocolate liquor and vegetable butters other than cocoa butter, and white and colored vegetable coatings that will withstand high temperatures and high relative humidities during the summer months, has increased the use of this important carbohydrate, sugar *(2)*.

Choice of Sweetener

The competitive relationship between sugar and corn sweeteners is actually in name only. In choosing the type of sweetener, a candy technologist is actually governed by the consumer, for it is only from repeat sales that confectioners are able to tell if they are satisfying consumers with their choice of carbohydrates in confections.

Candy technologists will continue to study the functions of carbohydrates in confections to give the highest quality possible. All they seek is good will. "Good will is the disposition of a pleased customer to return to the place where he has been well treated."—U. S. Supreme Court.

Literature Cited

(1) Alikonis, J. J., *Confectioner's J. and Mfg. Confectioner*, **78** (June 1952).
(2) Alikonis, J. J., Lawford, H., and Kalustian, P., "Role of Hard Butters in Chocolate-Type Coatings," N. Y. Section, American Association of Candy Technologists, February 1953.
(3) Cameron, A. T., "Taste Sense and Relative Sweetness of Sugars and Other Sweet Substances," New York, Sugar Research Foundation, *Rept.* **9** (December 1947).
(4) Ciccone, V. R., "Cooking Starch Jellies Continuously," 6th Production Conference, Pennsylvania Manufacturing Confectioners Association, April 1952.
(5) Dahlberg, A. C., and Penczek, E. S., N. Y. State Agr. Expt. Sta., *Bull.* **258** (1941).

(6) Farrell, K. E., and Alikonis, J. J., *Food Technol.*, **7**, 288–90 (July 1951).
(7) *Food Eng.*, **25**, 56–8 (February 1953).
(8) Jordan, S., "Confectionery Problems," Chicago, Ill., National Confectioners Association, 1930.
(9) King, J. A., "Art of Candy Making *vs.* Science," 4th Production Conference, Pennsylvania Manufacturing Confectioners Association, April 1950.
(10) Koorman, J., *Confectioner's J.*, **78**, 44 (March 1952).
(11) Koorman, J., "What Is Corn Syrup," *Confectionery and Ice Cream World* (June 1949).
(12) Krno, J. M., "Dextrose and Its Characteristics," 4th Production Conference, Pennsylvania Manufacturing Confectioners Association, April 1950.
(13) Lang, L., "The Role of Sugar in Candy Making," 3rd Production Conference, Pennsylvania Manufacturing Confectioners Association, May 1949.
(14) Martin, L. F., "NCA Candy Research Reports," New Orleans, La., Southern Regional Research Laboratory, 1951.
(15) Meeker, E. W., *Food Technol.*, **4**, No. 9, 361–5 (May 1950).
(16) Meeker, E. W., *Mfg. Confectioner*, **31** (February 1951).
(17) Pratt, C. D., "Sorbitol and Emulsifiers in Candy," Boston Section, American Association of Candy Technologists, November 1952.
(18) Slater, L. E., *Food Eng.*, **24**, 62 (July 1952).
(19) U. S. Dept. Agr., *Bull.* **48** (June 1951).

RECEIVED May 20, 1953.

Sugar in Confectionery

L. F. MARTIN

Sugarcane Products Division, Southern Regional Research Laboratory, New Orleans, La.

Problems in the use of sugars in confectionery have not received attention commensurate with their importance and the volume of sugars used. Review of recent advances in explaining the reactions of sugars in candymaking emphasizes the need for applying the methods and results of this research to the candymaker's art. Highly concentrated solutions of sugars at elevated temperatures undergo complex transformations, some initial products of which have been identified. The equally important reactions of sugars with proteins have been the subject of extensive recent research. The accumulated results of this work can be used to improve the quality of confectionery products. Formation of sugar anhydrides and caramelization products can account for many production difficulties and shortcomings of improperly processed hard candies. Determination of the extent to which these reactions proceed in practical hard candy manufacture, or can be controlled by variables in processing conditions, offers an opportunity for profitable research. Caramels are an example of candies in which it is desirable to accelerate as well as control the complex degradation reactions of sugars with proteins. They have been perfected as far as possible by rule of thumb methods, and any improvement must be based upon facts disclosed by fundamental research.

The confectionery industry uses approximately a billion and a half pounds of sugar and three quarters of a billion pounds of corn sirup and dextrose annually to produce candies and chocolate goods. In volume, it is the second largest industrial customer of the industries producing these sugars. A wide variety of other ingredients are combined with the sugars to produce 85 or 90 different items, but the chemistry of candy is primarily sugar chemistry. The reactions of sugars in various processes of confectionery manufacture largely determine the quality of the products, either by changes produced in the sugars themselves or by their reactions with the other ingredients. Recent research on the mechanism and products of these reactions applicable to candy problems has not received attention commensurate with its importance; therefore a review of pertinent results and their applications should serve a useful purpose. This review deals largely with progress in the chemistry of sugars rather than studies of the candymaking processes themselves. It is intended to point out opportunities for research and possibilities for using sugar more efficiently in candy production.

A basis for systematic review of this subject is provided by Table I, which shows the large total percentages of sugars entering into the composition of the more important types of candy, and the wide range of variation in the proportions used. It is evident that the chemistry of these primary ingredients of the candies must have the greatest importance in their formulation and processing. The table also gives the approximate range of cooking temperatures and final moisture contents for each type of candy. The most important reactions to be considered are those of sugars subjected to relatively high temperatures in extremely concentrated solutions, particularly hard candies, caramel, fudge, and "short" or grained marshmallows. In all but plain hard candy and brittle, reactions of the sugars with other ingredients under these conditions are also important. Jellies and soft marshmallows are made with less drastic modification of the sugars, as their consistency or "body" is provided by starch, pectin, or gelatin.

Action of Heat upon Sugars

The reactions of sugars in solution at high concentrations have received less attention, but useful deductions can be made by interpolating between observations made in the extensive research on dilute solutions and the dry sugars. Caramelization of sucrose has been studied by numerous investigators *(25)*. The stepwise dehydration of dry sucrose to "caramelan," "caramelen," and "caramelin" proposed by Gélis *(8)* has been supplanted by later work establishing the mechanism and products of at least the initial stages of reaction. Pictet and his coworkers *(19, 21)* identified the initial decomposition products, glucose and levulosan, formation of which precedes elimination of water in the reactions. Evidence for this reaction was obtained by Gélis *(7)*, but he was unable to separate and identify the products.

$$C_{12}H_{22}O_{11} \longrightarrow C_6H_{12}O_6 + C_6H_{10}O_5$$
$$\text{Sucrose} \qquad \text{Glucose} \quad \text{Levulosan}$$

The glucose formed is subsequently converted to glucosan, with evolution of the first molecule of water.

$$C_6H_{12}O_6 \longrightarrow C_6H_{10}O_5 + H_2O$$
$$\text{Glucose} \qquad \text{Glucosan}$$

Even if the water is rapidly removed by vacuum, at this stage the reactions will inevitably tend to follow the course of those occurring in very concentrated solution. The sugar anhydrides readily form dimers, levulosan in particular being converted into diheterolevulosans. It is now known that similar products are formed by refluxing fructose in 80% solution *(24)*. The reactions of caramelization cannot be limited to the initial stages without simultaneous further dehydration and polymerization, as well as extensive degradation to produce hydroxymethylfurfural *(13)*.

Table I. **Range of Cooking Temperatures, Moisture Contents, and Proportions of Sugars of Principal Types of Candy**

Candy	Final Cooking Temp. Range, ° F.	Final Moisture Content Range, %	Sugar Ingredients[a] Range, %			Principal Other Ingredients[a]	
			Sucrose	Invert	Corn sirup solids	Ingredient	Range, %
Hard							
Plain	275–338	1.0–1.5	40–100	0–10	0–60	......	......
Butterscotch	240–265	1.5–2.0	40–65	....	35–60	Butter	1–7
Brittle	290–295	1.0–1.5	25–55	....	20–50	......	......
Creams							
Fondant	235–244	10.0–11.5	85–100	5–10	0–10	Starch	0–1
Cast	235–245	9.5–10.5	65–75	....	25–40	Egg albumen	0–0.05
Butter	235–247	9.5–11.0	50–65	....	25–40	Butter	1–15
Fudge	240–250	8.0–10.5	30–70	0–17	12–40	{ Milk solids Fat	5–15 1–5
Caramel	240–265	8.0–11.5	0–50	0–15	0–50	{ Milk solids Fat	15–25 0–10
Nougat	255–270	8.0–8.5	20–50	0–15[b]	30–60	Fat	0–5
Marshmallow							
Grained	240–245	12.0–14.0	50–78	0–5	15–40	Gelatin	1.5–3
Soft	225–230	15.0–18.0	25–54	0–10	40–60	Gelatin	2–5
Jellies							
Starch	230–235	14.5–18.0	25–60	0–10	28–65	Starch	7–12
Pectin	220–230	18.0–22.0	40–65	....	30–48	Pectin	1.5–4

[a] Adapted from *(12)*.
[b] Honey is often used.

A key to the initial reactions of sugars in solution over a wide range of conditions is provided by the Lobry deBruyn–van Ekenstein transformation *(16)*. Solutions of pure sucrose rapidly become acid upon heating, and in candymaking the inversion thus produced is usually accelerated by addition of acid, or of invert sugar and glucose. Whatever candy formula is used, a system is established in which glucose and fructose are present, and they tend toward equilibrium with each other and with mannose. This interconvertibility of the three simple sugars, first discovered by Lobry deBruyn and van Ekenstein, proceeds to an extent governed by the concentration, pH, temperature, and ionic catalysts present. It is catalyzed particularly by alkalies, but takes place over a wide pH range. Maximum stability prevails for both glucose (23) and fructose (17) at a pH of approximately 3.0 to 3.3; in more strongly acid solutions the principal reaction is anhydride formation at the expense of the sugars (17, 24). Glucose can be transformed into fructose at pH 6.4 to 6.6 in the presence of phosphates and certain other salts (6).

Under all conditions of pH, concentration, and temperature that have been studied, the transformation of the reducing sugars by the rearrangement is accompanied by the formation of sugar anhydrides, particularly from fructose, which is the least stable of these hexoses. Gottfried and Benjamin (9) investigated the kinetics of this reaction under widely varying conditions and showed that the maximum yield of fructose obtainable from glucose is 21%, with the simultaneous formation of 8.5% of anhydrides and other unfermentable derivatives, and 3% of sugar acids. At extremely high concentrations or with only traces of water present, these reactions approach the changes associated with caramelization. Colored polymers are invariably formed, becoming the predominant products at the maximum concentrations in the presence of acids or salts.

Applications to Hard Candy Production Problems

Plain hard candies are regarded as the simplest type, as they are made entirely of sugars with added flavor, color, and sufficient acid to cause some inversion. They provide the best example of the importance of the chemistry of sugars in candy-making in the absence of complex reactions with other ingredients. Their production requires cooking temperatures above 300° F., or 270° F. with application of vacuum, to reduce the moisture content to approximately 1.5%. Substantial amounts of corn sirup, which is a complex mixture of glucose with varying proportions of maltose and dextrin, are generally added. Consideration of the reactions at high concentrations and temperatures described in the previous section makes it evident that the final chemical composition of hard candy may be as complex as that of any other type. Controlling color formation, preventing crystallization or graining, and minimizing the hygroscopicity or stickiness of the finished product are the principal problems in practical manufacture of hard candies.

Desirable properties in finished hard candy are known to depend upon the quality of the sugar used in its production. The "candy test" has been used by candymakers for control of sugar quality (2, 3). A hard candy is made from definite weights of sugar and pure water, cooked under carefully controlled conditions, and poured on a polished sheet of copper. The candy disk is observed to determine color and time of crystallization, as well as its tendency to become sticky. It serves to determine the "strength" of the sugar, which can be defined best as its resistance to inversion and further decomposition. Differences in quality and strength of sugars determined by this empirical test are attributable to the presence of very small traces of impurities (2). Addition of as little as 0.001% soda is known to have a strengthening effect in shortening the time required for crystallization. Water quality is particularly important in hard candy production for this reason, as traces of minerals in water may offset the advantage of using strong, or high quality, sugar.

The caramelization test of Pucherna (22) is another empirical test of sugar quality that has been modified recently by Kalyanasundaram and Rao (14). The latter workers improved the test by heating the sugar in glycerol, which permitted them to study the effect of additions of 0.1% of various salts and acids. Sodium chloride had the least effect in forming color, without measurable destruction of sugars, although it produced extensive inversion. Potassium chloride destroyed or transformed 4.2% of the total sugars without producing any color or causing measurable inversion. As would be expected, ammonium salts at this concentration caused the inversion of nearly all of the sucrose and destruction of a major proportion of the invert sugars formed, with maximum coloration.

Although the candy test is applied and interpreted as a measure of resistance to inversion, the strongest sugars being those which crystallize in the shortest time after cooling, crystallization is not desirable in hard candies. Corn sirup is usually added in sufficient amounts, or organic acids are used in pure sugar hard candies to produce enough inversion, so that crystallization is prevented almost indefinitely under proper packaging and storage conditions. It is unlikely that these measures would suffice for the purpose, were it not for the further decomposition of sucrose and its inversion products to form an extremely complex mixture. Sucrose, glucose, and fructose or their mixtures are not extremely hygroscopic, but the anhydrides of these sugars are very much so. Small percentages will account for the excessive tendency of hard candy to become sticky upon exposure to humid atmospheres. The maltose present in corn sirup also yields an anhydride on heating (20), the properties of which have not been investigated as thoroughly as those of glucosan and fructosan. Methods are now available (24) by which these complex sugar decomposition products may be separated quantitatively from mixtures with unchanged sugars. Their application to the analysis of hard candies produced under different conditions should make it possible to determine the extent to which such substances are formed and affect the quality of the product, thus providing the information necessary for rational improvement of processing methods and composition of the finished candies.

Production of hard candy with desirable color has been greatly facilitated by the wide adoption of vacuum cooking equipment with steam jacket heating. This makes possible more uniform heating and a reduction of about 30° F. in maximum cooking temperatures. Recent experimental development of high speed, continuous cooking equipment promises further improvement by shortening the time of heating necessary. Such improvements also make it possible to control inversion and decomposition of sugars to the anhydrides and degradation products which polymerize to form the colored substances associated with caramelization. In order to take full advantage of the precise control of process conditions permitted by this modern equipment, it will be necessary to determine the exact nature and extent of the chemical changes taking place and the precise conditions favorable to the formation of desirable final reaction products.

Reactions of Sugars with Other Ingredients

The chemical transformations of sugars are less important in candy which does not require cooking to temperatures and final concentrations as high as those necessary to produce hard candies. In most of the other candies listed in Table I, reactions of the other ingredients with the sugars assume greater importance. Initial products of sugar transformation are diverted to reactions involving these other compounds. The chemistry of certain side reactions has been studied extensively in recent years with some success in arriving at an understanding of the initial changes involved. Much of this research is directly applicable to the production of fudge, caramel, and other types of candies made with fats, proteins, starch, and other nonsugars. While details of these more complex chemical transformations are beyond the scope of this review, some of the obvious applications of recently acquired knowledge of the subject to improvements in candy production are worth noting.

The most important reactions in the production of candies made with milk are those of the sugars with the milk protein. In addition to the casein, milk introduces another sugar, lactose, which may enter into reactions similar to those of sucrose, increasing the number and complexity of the reaction products. Research up to 1951, and conclusions drawn from its results regarding the mechanism of the sugar-protein reaction, are summarized by Danehy and Pigman (4). In this case also, only the initial reaction steps have been investigated with any success; because these reactions are more complex, their mechanism and initial products have not been established with as much certainty as those of the decomposition of the sugars alone. More progress has been made in determining the course of the reaction of simple sugars such as glucose with individual amino acids instead of complex proteins (5). In the presence of amino compounds, transformation of the sugars is directed along a different path as a result of condensation of sugar-aldehyde and amino groups, followed by a rearrangement first described by Amadori (1), but only recently indicated to be of importance to an explanation of nonenzymatic browning in sugar-protein systems *(10, 11)*.

In spite of the complexity of these reactions, whatever has been learned of their chemistry and of the conditions that accelerate or retard them will be directly applicable in the production of the very large volume of candies in which proteins are used. Theories of these reactions and proposed mechanisms are still open to question, but the large amount of experimental work being done to prove or disprove the theories has already provided very many empirical observations that can be used to devise better candy cooking procedures.

Applications to Caramel Production

These complex reactions and further degradation of the initial products are not always undesirable. Caramels provide the principal example of candies in which such changes are deliberately produced, their flavor and color being dependent upon the interaction of sugars and proteins. The object is to develop the most attractive color and desirable flavor, while retaining suitable consistency or texture. There is considerable latitude for variation in reaction conditions to produce the various grades of caramels, ranging from the short, grained types approaching the consistency of fudge to very smooth and "chewy" candies. In any case, top quality products are obtained only through control of the extent of reaction, which is still attained by trial and error methods and subjective examination of the results. Attempts to use the latest equipment for high speed, continuous cooking by which the heating time is greatly reduced have not been successful in producing caramels, because reaction does not proceed sufficiently far to develop typical caramel flavor and color.

From leads obtained by experiments in simple systems containing pairs of individual sugars and amino acids, the investigation of reactions of proteins with more complex sugar mixtures has developed considerable evidence that condensation of sugar aldehyde groups with basic nitrogen of protein provided by lysine, arginine, and other basic amino acids is involved. This is currently the best available explanation of at least the initial stages of the process. The reaction of glucose with casein, the protein of milk, has been studied by Patton *et al. (18)*, who found considerable decrease in the amounts of lysine and arginine in the hydrolyzates obtained after reaction under their conditions. Reaction of casein with a variety of sugars was found by Lewis and Lea *(15)* to result in the loss of variable amounts of amino nitrogen, depending upon the particular sugar used. This immediately suggests a basis for the addition of particular sugars to caramel where it is desired to promote more extensive reaction, and formulation entirely with other less reactive sugars when the quality desired calls for minimum alteration of the ingredients.

Literature Cited

(1) Amadori, M., *Atti accad. nazl. Lincei*, [6] **9**, 68 (1929).
(2) Ambler, J. A., *Mfg. Confectioner*, **7**, No. 1, 17 (1927).
(3) Ambler, J. A., and Byall, S., *Ind. Eng. Chem., Anal. Ed.*, **7**, 168 (1935).
(4) Danehy, J. P., and Pigman, W. W., *Advances in Food Research*, **3**, 241 (1951).
(5) *Ibid.*, p. 248.
(6) Englis, D. T., and Hanahan, D. J., *J. Am. Chem. Soc.*, **67**, 51 (1945).
(7) Gélis, A., *Ann. chim.*, [3] **57**, 234 (1859).
(8) Gélis, A., *Compt. rend.*, **45**, 590 (1857).
(9) Gottfried, J. B., and Benjamin, D. G., *Ind. Eng. Chem.*, **44**, 141 (1952).
(10) Gottschalk, A., *Biochem. J.*, **52**, 455 (1952).
(11) Hodge, J. E., and Rist, C. E., *J. Am. Chem. Soc.*, **75**, 316 (1953).
(12) Jordan, S., and Langwill, K. E., "Confectionery Analysis and Composition," Chicago, Manufacturing Confectioner, 1946.
(13) Joszt, A., and Molinski, S., *Biochem. Z.*, **282**, 269 (1935).
(14) Kalyanasundaram, A., and Rao, D. L. N., *Sugar*, **46**, No. 3, 40 (1951).
(15) Lewis, V. M., and Lea, C. H., *Biochim. et Biophys. Acta*, **4**, 532 (1950).
(16) Lobry deBruyn, C. A., and van Ekenstein, W. A., *Rec. trav. chim.*, **14**, 156, 203 (1895); **15**, 92 (1896).
(17) Mathews, J. A., and Jackson, R. F., *Bur. Standards J. Research*, **11**, 619 (1933).
(18) Patton, A. R., Hill, E. G., and Foreman, E. M., *Science*, **107**, 623 (1948).
(19) Pictet, A., and Andrianoff, N., *Helv. Chim. Acta*, **7**, 703 (1924).

(20) Pictet, A., and Marfort, A., *Ibid.*, **6**, 129 (1923).
(21) Pictet, A., and Stricker, P., *Ibid.*, **7**, 708 (1924).
(22) Pucherna, J., *Z. Zuckerind. Cechoslovak Rep.*, **55**, 14 (1930).
(23) Singh, B., Dean, G. R., and Cantor, S., *J. Am. Chem. Soc.*, **70**, 517 (1948).
(24) Wolfrom, M. L., and Blair, M. G., *Ibid.*, **70**, 2406 (1948).
(25) Zerban, F. W., "Color Problem in Sucrose Manufacture," Sugar Research Foundation, *Technol. Rept. Ser.*, **2**, 2–5, Bibl. refs. 2–20 (1947).

RECEIVED July 10, 1953.

Sugar and Other Carbohydrates in Carbonated Beverages

CLAUDE GORTATOWSKY

The Coca-Cola Co., Atlanta, Ga.

In 1952 carbonated beverages accounted for about 13.1% of the total sucrose used in the United States and for a considerable tonnage of dextrose. The function of the carbohydrate is that of a sweetener and provider of body. Dextrose is employed by 27% of soft drink bottlers in a 5 to 45% mix with sucrose. An acidulated beverage showed, after 25 days' standing, 50% of its sugar as sucrose and 50% as invert, an indication that at the time of consumption an acidulated beverage carries a mixture of sucrose and invert, with a preponderance of sucrose. A committee set up under the sponsorship of the American Bottlers of Carbonated Beverages has proposed tolerances for color, ash, insolubles, bacteria, yeasts, molds, foreign odor, and taste.

Statistics on the utilization of sucrose in the carbonated beverage industry show an estimated 1,014,777 short tons for 1952 or 12.5% of the total estimated United States usage, 8,104,000 tons. There are consumed, in addition, an estimated 46,238 tons of dextrose.

The chief function of sucrose and other carbohydrates in carbonated beverages is as a sweetener. They also supply body to the beverage, often referred to as mouth feel. Sucrose serves both these purposes admirably. Another sweetener, dextrose, is being used in mixture with sucrose. During the war the use of dextrose in carbonated beverages received its chief impetus, and though usage figures have fallen below those of the war years, the current usage is many times prewar.

Interviews have shown that dextrose is rarely used alone, with the notable exception of its use in bread (7). Approximately 27% of the soft drink manufacturers interviewed reported they used dextrose in combination with sucrose. Fruit drinks, other than lemon-lime, particularly grape flavored, and root beer, were the relatively more important soft drinks where dextrose was used. A larger percentage of bottlers of root beer used dextrose as compared to bottlers of fruit flavors. Nearly one third of root beer bottlers, one fifth of those producing lemon-lime combinations, and 18% of ginger ale bottlers used dextrose, which ranged from 5 to 45% of the total sweetening agent; the greatest number preferred to limit it to 12 to 25%. Corn sirup was being used by only one manufacturer of carbonated beverages.

Sucrose Inversion

In carbonated beverages of the acidulated type, which constitute the bulk and have a pH range of 2.5 to 3.5, sucrose undergoes inversion; if the beverage is left to stand sufficiently long or at a sufficiently high temperature, complete inversion takes place. However, invert determinations made on a bottled beverage, at pH 2.5, show that sucrose is 50% inverted in 25 days at room temperature and completely inverted in 230 days. It is indicated, therefore, that a carbonated beverage of the acidulated type has, when consumed, more than 50% sucrose and less than 50% invert. On the assumption that the average age of the beverage is 2 weeks, the sucrose, based on the determinations referred to, was 63% and the invert 37% of total sugars. That is predicated on the use of 100% sucrose as the sweetening agent in a sirup bottled immediately after manufacture.

Frequently, for practical reasons, the sirup cannot be bottled promptly after it is produced and the invert sugar content will therefore be higher in the resulting bottled drink. Thus, a sirup was bottled after 7 days' standing and the bottled beverage held for 14 days, at which time it contained 80.6% of its sugar as invert and 19.4% as sucrose.

The question arises—what happens to beverage characteristics on inversion of the sucrose? Dahlberg reported that at a 10% sucrose concentration (a good average figure for sucrose content of carbonated beverages), dextrose had a sweetness of 79% with respect to sucrose and levulose and 115% with respect to sucrose *(2)*. Calculating resultant sweetness of a 10% sucrose solution converted to invert sugar as based on equivalent dextrose sweetness, we get from Dahlberg's figures:

> Original sucrose (10%) as dextrose = 12.7%
> 10% sucrose inverted = 5.26% dextrose + 7.68% (dextrose equivalent of levulose) = 12.94% divided by 12.7% = 102% based on original sweetness

Cameron reported that in 10% solution of sucrose as a basis of comparison, dextrose had a relative sweetness of 68.5% and fructose or levulose was 20.5% sweeter than sucrose *(1)*. The most widely accepted sweetness for anhydrous dextrose is 70 to 75% that of sucrose. Calculating resultant sweetness of a 10% sucrose solution converted to invert sugar based on equivalent dextrose sweetness, we get from Cameron's figures:

> Original sucrose (10%) as dextrose = 14.6%
> 10% sucrose inverted = 5.26% dextrose + 9.38% (dextrose equivalent of the levulose) = 14.64% and 14.64% divided by 14.6% = 100%

Cameron therefore confirmed Dahlberg's findings with respect to relative sweetness of invert sugar resulting from and as compared to sucrose in 10% concentration. He reported further a slight loss in sweetness on inversion of 1 to 10% solutions of sucrose and a small gain from 10 to 20% sucrose.

Dahlberg's work was done at 70° F. and Cameron followed Dahlberg's procedures. It has been reported that lowering temperatures in effect decreases intensities of sweetness. Sourness before sweetness increases the tongue's sensitivity for sweetness and it has been suggested that other stimuli, even though tasteless, may modify sweetness. Soft drinks are consumed cold; the acidulant contributes tartness or sourness and there are flavors that may affect sensitivity to sweetness. Because sweetness evaluations are average figures derived from a number of subjective tests, it cannot be proved that sweetness in every carbonated beverage is the same before and after inversion. But practically, such is probably the case. No change in sweetness of 10% sucrose solutions as a result of inversion was reported in water or carbonated beverages *(3)*.

Requirements of Sweetening Agent

The sweetening agent of a carbonated beverage is under test until the beverage has been consumed. It is subjected to the delicate judgment of the senses of taste and of odor as well as touch, more commonly referred to as mouth feel. The sweetening agent must be of such a quality that the beverage will remain clear and not cloud or become turbid (must be color-free within tolerances), will not throw down insolubles, will be free from foreign taste and odors, and will have proper body.

Quoting from Spencer and Meade *(6)* "The impurities, particularly color and colloids causing haziness in solution, are of great importance in many processes such as making of fine candies, condensed milk, water-white sirups, and soft drinks. For most practical purposes the measures of the impurities present is the color and appearance of a solution of the sugar . . ."

Many are familiar with the Sugar Industry Bottlers' Committee consisting of 14 members: 6 from the soft drink industry, 4 domestic cane refiners, 3 beet sugar processors, 1 off-shore cane refiner. Through its subcommittee that committee has promulgated tentative standards for so-called bottler's sugar. A color tolerance of 35.0 determined by the Holven-Gillett-Meade method has been proposed for color. Such a figure is very liberal and could be lowered to 25.0, allowing proper tolerance, without causing hardship. The committee has proposed that the sugar be free from "obviously objectionable taste or odor" in dry form or in 10% solution. The same should be true in 50% solution.

The sugar should be free from insoluble particles. The committee has proposed a tolerance based on approximately 2 p.p.m., to be indicated by means of a filter disk 1⅛ inches in diameter, on which has been deposited carbon to duplicate the appearance of 2 p.p.m. pan scale. The actual test is conducted by dissolving 300 grams of sugar in distilled water and filtering through a 1⅛-inch disk of Whatman No. 40 or comparable filter paper.

A figure of 0.015% has been proposed by the committee as the tolerance for ash. The determination is to be made by the conductivity method, the specific conductance in micro-ohms multiplied by a factor of 0.0005. This method is based on ashing, resulfation, and deducting 10%. Some users do not look favorably upon the suggested tolerance of 0.015% and feel the figure should be not more than 0.01%.

Dissolved in distilled water, the sugar should give a solution free from turbidity in 10 or 50% concentration. Pending completion of work in progress, a tolerance for turbidity will be given further consideration. Needed are a reliable instrument and simple method for making routine turbidity measurements. Visual methods are now employed. Turbidity should be absent in neutral or acid solutions.

The sugar should be free from any traces of antifoaming agents. Presence of these may kill the bead or collar on a carbonated beverage, making it appear flat.

The sugar should be free from floc-producing substances. A 50 to 55 Brix solution at pH 1.5 should remain free from floc 10 days or longer, the acidity being attributed to any of the food acids. The same should be true in 10% solutions.

Bacteria

It has been said that refined sugar should be produced under rigid bacterial control; and that the necessity for these measures is twofold—insurance of the keeping quality of the sugar itself during prolonged storage, and elimination of the possibility of sugar's contributing to certain definite spoilage problems in the products in which it is used (5). Microbial flora of cane sugars are, according to Owen, bacteria (principally spores of mesophilic and thermophilic types), yeast (principally of the nonsporulating or torula group), and mold fungi (5).

Bacterial contamination can cause cloudiness in the drink; yeasts and molds can cause fermentation and formation of acids resulting in souring, and the appearance of foreign bodies; all make the beverage unacceptable and unfit for sale.

The Sugar Industry Bottlers' Committee, through its subcommittee, has set up the following tentative bacterial standards for "bottler's" sugar. These standards apply to sugar as produced, immediately prior to packing.

Mesophilic bacteria	Not more than 200 per 10 grams
Yeasts	Not more than 10 per 10 grams
Molds	Not more than 10 per 10 grams

Liquid Sugar

Liquid sugar may be considered as 100% sucrose No. 1 and 50/50 invert sucrose sirup.

No. 1 sucrose sirup is made by (1) boiling down No. 1 sugar liquors that have been treated with bone char and vegetable carbon, and (2) by dissolving granulated sugar, with subsequent sterilization and filtration. Dissolved granulated sugar will have a low ash content and will comply with standards for granulated sugar. Liquid sugar in which the sucrose has not been boiled out in a pan will have a higher ash content. Both are employed in the carbonated beverage industry. Ash content of the latter type may be in the order of 0.05% on a wet basis. Some object to such ash content per se. Others feel such quantity of ash to be important from a qualitative viewpoint and reason that a most important quality is low content of organic nonsugars, particularly acids of the fatty acid series.   These acids are set free from their salts on acidification and, if present even in minute quantity, may impart an objectionable character to a sirup or beverage.   Butyric acid is detectable in air to the extent of 1 part in 12,500,000 by weight (4).   A typical No. 1 liquid sugar shows 67.0% solids, invert sugar 0.2%, and ash 0.05%, pH 6.8, color water white, turbidity none; 50% invert sirups: 76.0% solids, ash 0.05% or as low as 0.01% if made by ion exchange, pH 5.0, straw color developed through inversion, turbidity none.   These sirups should be free from bacteria, yeast, and molds.

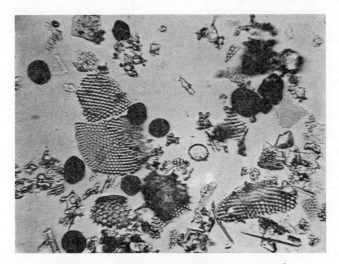

Figure 1.   Stained Starch Granules Recovered

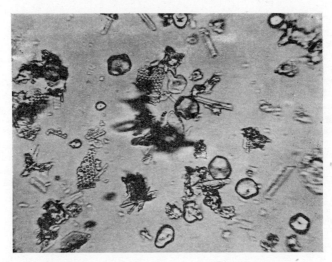

Figure 2.   Unstained Starch Granules Recovered

As a trace contaminant in sugar, also found occasionally in the carbonated beverage, starch granules settle on the bottom of the bottle (Figures 1 and 2). Starch granules have been found in numerous cases on examination of sugars. Their origin remains in doubt.

Literature Cited

(1) Cameron, A. T., *Can. J. Research*, **22E**, 45–63 (1944).
(2) Dahlberg, H. W., and Penczek, E. S., N. Y. State Agr. Expt. Sta., *Tech. Bull.* **258** (April 1941).
(3) Miller, W. T., *Food Packer*, **27**, No. 13, 50, 54 (1946).
(4) Moncrief, R. W., "The Chemical Senses," p. 80, New York, John Wiley & Sons, 1946.
(5) Owen, W. L., and Mobley, R. L., *Facts About Sugar*, **30**, No. 12, 451–2 (1935).
(6) Spencer, E. F., and Meade, G. P., "Cane Sugar Handbook," p. 557, New York, John Wiley & Sons, 1945.
(7) U. S. Dept. Agr., *Inform. Bull.* **48** (1951).

RECEIVED August 27, 1953.

Sugars in the Canning of Fruits And Vegetables

P. W. ALSTON

Spreckels Sugar Co., 2 Pine St., San Francisco, Calif.

For many years only cane sugar was considered as the essential sugar for canning. Improved methods of beet sugar refining made beet sugar as satisfactory as cane sugar. During the second quarter of this century starch conversion products became an economic factor in the canning of fruits and vegetables. Sugar and starch conversion products have different physical and physiological properties and cannot be indiscriminately substituted one for the other. It appears that there is much need for sound research to determine the proper medium for each product on a basis of quality as well as of economics.

Few developments during the past century have added more to the increase in the standard of living than the contribution of the art and science of canning. Canning may be defined as the method of food preserving in a permanently sealed container.

The invention of the canning process by Nicholas Appert and the extraction of sugar from the beet by Delessert, both in France during the Napoleonic Wars, were the results of food shortages caused by blockade.

In the early descriptions of methods of preservation of fruits, water alone was used to complete the filling of the container—usually a bottle. The high price of sugar prompted the offering of awards for alternative methods of food preservation. Five guineas were voted to Thomas Saddington for a cheap method of preserving fruit without sugar for use as sea stores. He advised the consumer to use part of the first liquor poured off for flavoring pastry and to cook the remainder of the pack with a little sugar. Increase in sugar production soon brought the price down where the sugar could be added during the canning process. In 1810 sugar sold in London for 60 shillings per pound. A skilled artisan, such as a carpenter, received 30 shillings per week. Today, a week's work would buy 1200 pounds of sugar.

Sucrose was the only suitable sugar available until the second quarter of the present century. In the first quarter of the century much time and temperature, which are both essential in their proper usage, were wasted in a controversy over the relative merits of beet and cane sugar as preserving media. This conflict died down with the development of the present methods of processing beets to granulated sugar. Today practically all canners in the West use both cane and beet sugars without distinction of quality differences. Just as World War II resulted in a new alignment of opposing political forces, the beet and cane sugar industries and the corn products industry now confront each other in a polemic discussion of the relative merits of the several media used in canning. Reading the literature on the subject of the preferred media leaves one as confused on the problem as he is likely to be after the campaign speeches in a national election. The chief impression to be obtained from such literature is that scientific conclusions are not the objective of much of the experimentation reported; rather the presentations are offered for purposes of propaganda.

Joselyn (1), reporting on the influence of the several media on flavor, texture, and color in the frozen food industry, states that there is too little technological information on the basis of which proper selection of sugar can be made. This remark can be made with greater emphasis in regard to the older art of canning. The newer industry of quick-freezing required intensive work in order to produce an acceptable product that could compete with the canned goods. However, the older industry has not deemed it necessary to determine so precisely the effect of the several materials on the quality of the finished product. Economics has apparently been the principal stimulus to the use of mixtures of various sugar, so much so that the most voluminous information available is in the form of polemic argumentation before the federal agency concerned with what is printed on a label.

It would appear to any conscientious investigator of the question of sugars in canning that there is no one answer to all the packs. The fact that a pure sucrose sirup is best for one particular fruit does not indicate that it is the best for all. There is such a wide variety of original materials and a divergence of properties of packing media that it must be an axiom that concerted, concentrated, and honest research is necessary to develop the optimum for each particular case. The high income of the American customer makes him quality conscious.

An indication of the type of concerted research desired may be gathered from the few scattered reports available. The stability of certain pigments is affected by levulose and furfural. Mixtures of sugar are preferred for mild-flavored fruits to maintain a desired density without excessive sweetness which tends to mask the delicate flavor. In canned vegetables sugars are not desired as sweeteners, but are used in low concentration as a seasoner.

Industrial Uses

The industrial usage of sugars as sweeteners has increased twofold in the past 15 years, during which time the population has increased by 19%. This represents an increase in income and also a decrease in home preparation. This tendency is expected to grow with increase in income and population, which will ensure increased demand but will also cause more careful selection on the basis of quality. It seems that the problem here is not one of preference of one medium over the other, but what is the most suitable for each particular article of commerce. An increase in quality will result in an increase in demand, which, in turn, will benefit all parties concerned.

The present trend here in the West has been to all-liquid media. Beet and cane refiners supply liquid sugars to the canners as liquid sugar of the sucrose type. Starch conversion products in liquid form are supplied as corn sirup unmixed, high conversion corn sirup, and enzyme conversion corn sirup. The adoption of the liquid media in canning has been of immense economic advantage to the canning industry in California.

For the canning of certain types of nonacid fruits and vegetables the canning medium must be low in acid-forming thermophilic bacteria. Specifications issued by the National Canners' Association (2) are easily met by the competent producers of either medium.

Sulfur dioxide is undesirable if the concentration is above a few parts per million. The delicate color of the fruit is impaired and in some cases the metal of the can is corroded. Direct consumption cane sugars produced by the sulfitation process are usually undesirable for canning but the supply is limited to one section of the country. Sulfur dioxide is used in processing beet sugars but not as a bleach as was the practice earlier in the industry. It is now recognized that a few parts per million of sulfur dioxide added before the evaporation of the thin juice act as an inhibitor to the nonenzymatic browning and result in a sugar with less than 1 p.p.m. of sulfur. In the same manner the starch-conversion producers have been able to eliminate the objectionable hydroxymethyl furfural, an undesirable side reaction product in starch conversion.

Artificial Sweeteners

Some special dietetic packs are prepared with artificial sweeteners. Saccharin, benzoylsulfonic imide has been used since the eighties of the last century. More recently a new nonnutritive substance, sucaryl, the sodium salt of cyclohexanesulfonic acid, has been produced in two forms—Sucaryl sodium and Sucaryl calcium. The calcium type is proposed for those whose ordinary salt ration should be limited. The Food and Drug Administration treats these compounds in the manner prescribed for saccharin.

Two other artificial sweeteners, dulcin (sucrol) and P-400, are classed as poisonous substances and are forbidden for use in food preparation.

Literature Cited

(1) Joslyn, M. A., et al., *Food Technol.*, **3**, 8–14 (1949).
(2) Owen, W. L., "The Microbiology of Sugars, Syrups, and Molasses," p. 156, Minneapolis, Minn., Burgess Publishing Co., 1949.

RECEIVED September 3, 1953.

Sugars in the Baking Industry

SYLVAN EISENBERG[1]

Chemistry Department, University of San Francisco, San Francisco, Calif.

Bakery products fall into two main groups—yeast-raised and nonyeast-raised. Yeast-raised goods account for roughly 35% of the sugars used in the United States baking industry. The principal functions of sweeteners in yeast-raised products are to provide fermentability and sweetness; in nonyeast-raised products, sweetness. Sucrose and dextrose are the principal sugars purchased in highly refined form. Fructose, lactose, and maltose find their way into baked goods in the form of invert sirups or honey, milk, and malt sirups. These sugars differ in sweetness and in fermentability by yeast, as well as in physical properties such as hygroscopicity.

Bakery products may be classified into two main groups—those leavened by yeast and those by chemical or physical means. Yeast-raised goods include bread, rolls, sweet rolls, raised doughnuts, and crackers. Chemically leavened products include most cakes, cake doughnuts, pastries, and cookies. Foam-type cakes such as angel food and sponge cake are physically leavened by aeration. Pies represent a separate class.

Of the total estimated bakery production in 1947 (8) approximately 79% was yeast-leavened. In terms of sweetener consumption, however, yeast-raised goods accounted for only 35% of total sweeteners purchased by the baking industry, as computed from known quantities of the individual types of merchandise produced (8) and estimates of added sweetener content for each type as required by their formulations. This disproportionate relationship between raised goods produced and sweeteners used in their production merely reflects the fact that such goods are relatively unsweetened. Pertinent data are shown in Table I.

Table I. Estimated Amounts of Sweeteners Used for Baked Products in the United States in 1947

(Millions of pounds)

Products	Quantity (8)	Sweetener, %[a]	Sweetener Used (Estimated)
Yeast-raised goods, 79%			
Bread	10,500	3.7	390
Rolls	942	5.0	47
Sweet goods	612	9.0	55
Crackers	928	0.0	0
Total	12,982		492, 35%
Av.		3.8	
Nonyeast goods, 21%			
Doughnuts	400	20	80
Cakes and pastries	1,291	33	430
Cookies	1,111	25	280
Pies	601	20	120
Total	3,403		910, 65%
Av.		27	
Grand total	16,385		1402
Total sucrose used by U. S. baking industry, 1947 (8)			1139
Fraction of total sweetener represented by sucrose (5)			76 %
Total sweetener used by U. S. baking industry, 1947[b]			1486

[a] Estimated from trade formulas.
[b] Calculated (5, 8).

[1] Present address, Anresco, 693 Minna St., San Francisco, Calif.

Kinds of Sweetener Used

The three principal sweeteners purchased as such are sucrose, dextrose monohydrate, and corn sirup; sucrose represents about 75% of the total. Relatively small quantities of malt sirup, honey, invert sirup, and molasses find special use. Two other sugars occur in bakery products in significant quantities, though often overlooked. Lactose is introduced through nonfat dry milk, and must approach at least 10% of the total sucrose used. Maltose, other than that directly added as malt sirup, is invariably introduced by the influence of diastatic enzymes on flour starch. Quantities so introduced are difficult of estimate, but there are some data for yeast-raised goods. Rice (7) reports maltose content of bread on a dry basis, from which a value of 2.8% maltose can be calculated on a fresh bread basis. This figure is certainly of the right order as judged by the known diastatic activity of commercial bread flours, and it may be reasonably applied to yeast-raised goods in general. On this basis, maltose is believed to occur in baked goods to the extent of 32% of the total sucrose used by the United States baking industry—that is, in bread at least maltose is a principal sugar.

Sucrose. Sucrose is the sweetener purchased in largest amount, and, as might be expected, it is produced in many forms. Besides the several grades of brown or so-called soft sugars which have limited use in some icings and in dark variety goods, highly refined sucrose can be classified into three main groups—granulated, powdered, and liquid.

The grades of granulated sugar differ from each other in crystal size, which is controlled by processing and not by grinding of coarser material. The finer products find particular application in dry mixes and in cakes, the coarser for general use and for making cooked fondants.

Powdered sugar is produced by grinding granulated stock, and is available in several grades of different mesh. Uniformity of particle size within grade is important, particularly if the sugar is to be used in cream fillings or coatings, where coarse grains may produce grittiness. Of the several grades made, the finer are used in making cream fillings and coatings and in preparing uncooked fondants, the coarser for dusting fried goods such as doughnuts.

Liquid sugar, pure sucrose sirup, is relatively new to the baking industry, and use at present is confined to the larger bread manufacturing plants. Such plants, when they are located not too far from the source of supply, can take advantage of a price differential which more than offsets the cost of shipping contained water.

Dextrose Monohydrate. Second in importance only to sucrose, this sugar finds considerable use in the manufacture of raised goods, particularly bread, and in many plants it has displaced sucrose entirely. Other grades of dextrose, powdered and pulverized, are used in cakes, cookies, and icings. In such products up to 25% of the sucrose has been replaced with dextrose, particularly in chocolate cake (5, page 107). This provokes thought, inasmuch as chocolate cake and chocolate goods in general differ from others by being alkaline, and in them sucrose would undergo a minimum of inversion. De Boer (2) contends that interconversion between dextrose and levulose occurs in cakes in the presence of bicarbonate or carbonate leavening.

The combined solubility of two sugars is greater than that of either alone, though less than the sum of the two (5, page 56). This would apply possibly to the combined use of dextrose and sucrose in high sugar formulas, particularly as only part of the water in baked goods is available as solvent (6).

Other sugar products. Corn sirup and invert sirup find use in cakes, cookies, icings, and fillings, where their hygroscopic properties are useful in increasing the amounts of moisture retained. In icings these sirups also serve as stabilizers, in the sense that they retard crystal growth of other sugars. Honey and molasses are also used in the same way when their characteristic flavors are required. Nondiastatic malt sirup is used in raised goods and in some cookies, to supply fermentable sugars for the one, or to impart characteristic flavor, color, and moisture-retaining ability to the other.

Functions of Sweeteners

Nonyeast Goods. Here the most important function is to impart sweetness. This practically eliminates maltose and lactose from consideration. Sucrose and dextrose are the two most important sweeteners purchased.

Possibly next in importance is the ability of the sweetener to caramelize, imparting crust color. This is important in yeast-raised goods as well, and certainly in such products caramelization is a complex process affected by components other than sugars. In nonyeast goods sweetness controls, and discussion of crust color as a function of kind of sugar is academic, except for one application. In the making of piecrust nonfat dry milk is often used to impart color. Sucrose and dextrose allegedly encourage sogginess. Possibly lactose with its low solubility could be used to advantage, price permitting.

Sweeteners have other functions as well. They impart tenderness to cakes, thus enhancing their practical keeping quality. They increase the mobility of batters and the spread of cookies. The sirups increase moisture-retaining ability and retard crystallization of other sugars.

Yeast-Raised Goods. First in importance, but possibly second in some instances, is the ability of the sweetener to support fermentation, for only by this means is leavening achieved in this kind of merchandise. Second in importance, but possibly first in some instances, is the requirement of sweetness. In sweet goods at least, and also in bread as judged by present trends, sweetness is becoming more important. These functions must be served. Besides these, sweeteners contribute crust color and toasting quality. Here differences among various sweeteners, though important, have not been established with sufficient objectivity to warrant discussion. With regard to dextrose and sucrose, however, Barham and Johnson (1) report no practical difference. Sweeteners are also said to influence keeping quality, some sugars being more effective than others. Such differences again are not considered clearcut; so-called bread softeners are more effective.

Sweetness

Sweetness cannot yet be determined by chemical or instrumental means. Estimates of the relative sweetness of different substances remain dependent upon statistical treatment of data obtained subjectively by means of taste panels. Reduction of such results to practical application appears simple, but may indeed become illegitimate.

Cameron (3) and others (5, page 48) have found that sweetness is not a linear function of concentration, and that the relative sweetness of different sugars varies with the concentrations at which they are compared. Sweetness is influenced by temperature, pH, and the presence of other substances which need not themselves be sweeteners. For example, a 5% sucrose solution containing 2% urea was found to be equal in sweetness to a 3.1% sucrose solution. Thus, sweetness had been reduced 38% by the presence of urea.

Though these difficulties must be kept in mind, the quality of bakery products must be controlled, and bakers are prone to flirt with a single numerical "score" of quality. Doing this, they are also impressed with similar scores of the sweetness of different sugars. Such scores of relative sweetness, with sucrose arbitrarily assigned the value 100, are shown in Table II.

There is evidence that at high concentrations the less sweet sugars become relatively sweeter compared to sucrose and that synergistic effects among sugars act in the same direction. Thus, Cameron reports that invert sugar is sweeter than sucrose at concentrations above 10%, but not so sweet as sucrose at lower concentrations.

Fermentability

Of the monosaccharides involved in raised-goods manufacture, dextrose and levulose are fermentable. Of the disaccharides, sucrose and maltose are fermentable whereas lactose is not. This much is clear. Difficulties arise, however, when relative fermentability of different sugars is considered.

Answers to this question have been sought by two methods, measurement of carbon dioxide produced or determination of residual sugars. Each of these techniques has been applied either to fermenting doughs or to other fermenting media. If concern with fermentability is limited to leavening—and this is certainly justified—then only those methods that measure gas production in fermenting doughs are pertinent. And even with this limitation we have at least two possible criteria: efficiency of conversion to carbon dioxide and rate of conversion. Both factors may well vary and vary differently for different sugars, with modification of fermentation conditions and materials.

Table II.   Relative Sweetness of Sugars and Sugar Products

Levulose	140–175
Invert sugar	100–130
Sucrose	100
Dextrose, anhydrous	70–75
Dextrose monohydrate	60–75
Enzyme-converted corn sirup	60
Corn sirup, unmixed	30
Maltose	30
Lactose	15
Galactose (estimated from data of Cameron)	58

A promising attack on this problem has been made. In plotting instantaneous rates of gas production against fermentation time (4), a region of exponential decline of rate is found as sugar exhaustion approaches. In this region the fermentation functions as a first-order reaction, and one may write:

$$R_t = K(V_T - V_t)$$

in which R_t is the instantaneous rate at time t, K is the velocity constant, V_T is the total amount of gas to be expected from the sugars existent up to and included in the first-order range, and V_t is the total volume of gas produced up to time t.

Now if R_t is plotted against V_t, a linear region corresponding to the first-order range should be found, and this is done with varying degrees of success, depending upon the fermentation conditions and materials. The slope of this line is K. The V intercept is V_T. This intercept can be determined with almost analytical accuracy; the slope, with sufficient accuracy to justify statistical study.

With V_T determinable, it becomes possible to measure increments in volume of gas produced as a function of kind and quantity of sugar; hence the efficiency of conversion to carbon dioxide. K, other conditions the same, measures the rate at which the sugar is converted to carbon dioxide. Table III reports data for sucrose, dextrose, and maltose, where these sugars were added to the dough stage of an 80% sponge fermentation. Sponge times were 5.5 hours in one case and 24 hours in another. All figures are averages of two independent determinations. Volumes of carbon dioxide per gram of sugar have been adjusted to 30° C. and 760 mm. of mercury pressure, and to the anhydrous basis.

Table III.   Fermentability of Sucrose, Dextrose, and Maltose

V = volume of CO_2 per gram of sugar
K = specific reaction rate

	V			K		
Sugar	5½ hr.	24 hr.	Av.	5½ hr.	24 hr.	Av.
Sucrose	219	220	220±1	0.34	0.53	0.44±0.10
Dextrose	216	218	217±1	0.55	0.63	0.59±0.04
Maltose	200	208	204±4	0.55	0.63	0.59±0.04

These results establish that sucrose and dextrose in chemically equivalent quantities are converted to carbon dioxide with practically identical efficiencies, whereas maltose is some 10% less effective. Maltose and dextrose under the conditions investigated fermented with practically identical specific reaction rates. The induction period usually observed for maltose under sponge conditions of fermentation was not observed here. Sucrose appears to have a smaller K value. This indicates that the rate of conversion is somewhat lower than for the other sugars under the specified conditions, but the difference is too small to have practical significance.

It is possible that sugars other than those discussed may become important to the baking industry. Interest in the production of pure levulose continues. Lactose has dropped from about 60 to 14 cents per pound over the past 15 years, whereas the cost of nonfat dry milk has increased about 300%. If lactose were hydrolyzed, perhaps even in the dough, its fermentability would be increased from nil to 50% that of dextrose and it would become as sweet.

Literature Cited

(1) Barham, H. N., Jr., and Johnson, J. A., *Cereal Chem.*, **28**, 463 (1951).
(2) Boer, H. W. de, *Chem. Weekblad*, **47**, 269–74 (1951).
(3) Cameron, A. T., Sugar Research Foundation, *Sci. Rept. Ser.* **9** (1947).
(4) Eisenberg, S., *Cereal Chem.*, **17**, 430–47 (1940).
(5) Jones, P. E., and Thomason, F. G., Production & Marketing Administration, U. S. Dept. Agr., *Agr. Inform. Bull.* **48** (1951).
(6) Kuhlmann, A. G., and Golossowa, O. N., *Cereal Chem.*, **13**, 202 (1936).
(7) Rice, W., *Ibid.*, **15**, 672 (1938).
(8) U. S. Bur. Census, *Northwestern Miller*, Section 2, **245**, 113–15 (1951).

RECEIVED September 2, 1953.

Sugar and Other Carbohydrates in Meat Processing

H. R. KRAYBILL

American Meat Institute Foundation, and Department of Biochemistry,
The University of Chicago, Chicago, Ill.

Sugar is used in large quantities in meat curing, though its role in meat processing is not completely understood. Sugar is not present in most cured products in sufficient amounts to impart a sweet taste, but it may serve to soften the brashness of the salt. Bacon is an exception, as the sugar content of the fried bacon is high enough to impart a distinct sweet taste. Additional research is needed to determine the role of sugars in the development of cured meat flavor. Sugars play an important role in the curing by maintaining acid and reducing conditions favorable to good color development and retention. Under certain conditions reducing sugars are more effective than nonreducing sugars, but this difference is not due to the reducing sugar itself. The exact mechanism of the action of the sugars is not known. It may be dependent upon their utilization by microorganisms or the enzymatic systems of the meat tissues. Considerable quantities of starches and flours are used as binders in sausages and prepared meats. They serve to retain moisture throughout processing and storage of the product, and also may stabilize the emulsion of moisture, fat, and protein. Recent experiments indicate that the quality of beef and pork can be improved by ante-mortem feeding of sugar.

Accurate data on the amount of sugar used in the meat industry are not available, but estimates calculated from federal statistics on the amount of meat placed in cure indicate that the amount of sugar used is in the order of 50,000,000 to 60,000,000 pounds annually (25). Sugars are an important constituent of the curing ingredients used extensively in processing numerous types of cured sausages and bolognas, fresh pork sausage, hams, picnics, butts, bacon, corned beef, dried beef, canned spiced ham, luncheon meats, meat loaves, and chili con carne. The function of sugar in meat curing is not completely understood.

The purpose in curing meat is to prolong the keeping time and develop a desired flavor and color. Before adequate refrigeration was available, the primary purpose of curing meat was to prolong the keeping time. Today milder curing processes are used and many of the products must be kept under refrigeration. The development of extensive refrigeration facilities has made it possible to produce cured meats of higher palatability and greater acceptability. Thus, the primary emphasis has shifted to obtaining the desired flavor, tenderness and color.

Type of Sugar Used

Sucrose is the sugar used most extensively in curing meat. In experiments carried out by the Department of Scientific Research of the American Meat Institute on hams and bacon, no difference in the finished product or course of the cure was found when beet or cane sugar was used. Raisin sirup, honey, molasses, and different grades of refiners' sirups have been used to a limited extent in curing meats. Dextrose and corn sugar sirup are used in some meat products.

Because of a sugar shortage during World War I, Hoagland *(14)* carried out a series of experiments in a number of packing plants in which he compared granulated cane sugar, dextrose, cerelose, 70% corn sugar, and refiners' sirup for curing pork hams, bacon, and beef hams. Very little difference in quality was found in the pork hams cured with the different types of sugar. The quality of the bacon varied very little, except that the bacon cured with dextrose and cerelose browned too readily when fried. The beef hams cured with dextrose and cerelose were equal in quality to those cured with cane sugar. Those cured with 70% corn sugar and refiners' sirup were of inferior quality.

Lewis and Palmer *(20)* carried out extensive experiments comparing raisin sirup, cerelose, and sucrose in curing pork hams, bacon, beef tongues, and beef hams. They found no significant difference in quality of the pork hams. The bacon cured with corn sugar was superior to that cured with raisin sirup or cerelose. When a partially refined cane sugar (95% total sugar) was compared with refined cane sugar in curing hams, no significant differences were found in the quality of the hams.

Amount of Sugar in Cured Meats

Table I, based on Mighton's *(21)* studies, shows that the average sugar content of cured meat varies from 0.10% in canned corn beef to 2.32% in cooked bacon. Gross *(11)* later analyzed a number of commercial hams produced by six different packers and obtained on the open market. The average sugar contents were: tender smoked, 0.65%; ready-to-eat smoked, 0.74%; boned and rolled in transparent casings, 0.31%; and canned, 0.75%.

Table I. Sugar Content of Cured Meats
(Expressed as per cent sucrose)

Product	Maximum	Minimum	Average
Commercial boiled hams	0.71	0.26	0.55
Regular boned hams	0.64	0.25	0.42
Pork butts	0.89	0.18	0.60
Canned corn beef	0.20	0.03	0.10
Canned spiced ham	3.20	1.15	1.68
Picnics	0.47	0.07	0.28
Frankfurters	1.65	0.41	1.26
Bologna	1.88	0.19	1.11
Bacon, raw	0.93	0.30	0.72
Bacon, cooked	3.28	0.96	2.32

The amount of sugar actually incorporated in most cured meats is probably too low to impart a definitely sweet taste to the product, but it may serve to soften the brashness of the salt *(16, 17)*. Bacon represents an exception, as the cooked product contains a higher percentage of sugar, which imparts a sweet taste. When hams and pork butts were cured with and without sugar and subjected to taste panel tests *(1, 10, 24)*, the results indicated that sugar is not a dominant factor in the flavor of these products. Additional, more critical tests are needed to determine definitely what part, if any, the sugars play in development of the flavor of cured meats.

Effect of Sugar on Color

While carrying on a study on the use of cane sugar and corn sugar in the curing of pork butts, Lewis, Oesting, and Beach in 1936 *(19)* made an important discovery, which called attention to the probable function of sugars in the development and retention of the color of cured meats. They observed the color changes in laboratory samples of ground pork butts, one of which had been cured with cane sugar, a second with corn sugar, and a third with no sugar. In all other respects the treatment of the pork butts was identical. The color of the sample cured without sugar faded very rapidly, the sample cured with corn sugar held its color well, and the sample cured with cane sugar was intermediate in respect to color retention. The grinding of the samples, exposing a larger surface to diffuse light and oxygen of the air, resulted in an accelerated test of color stability. Previous observations had shown that light and oxygen were important factors in the fading of cured meat color.

To test the observation further, they ground pork butts that had been cured with sugar and divided the meat into four lots. To one portion cane sugar was added, to a second corn sugar, to a third honey, and the fourth was untreated. When observed under diffuse light in the laboratory, the untreated control faded very rapidly. The cane sugar sample faded also, but the two treated with honey and corn sugar retained their color. Lewis and coworkers then cut some of the pork butts and exposed the surfaces to air and diffuse light; the color of the butts cured with corn sugar held up better than either of the others. The butts cured without sugar faded on the cut surfaces.

In another experiment unsmoked pork butts cured without sugar were divided into three lots. One lot was soaked for an hour in plain water, a second in 10% cane sugar, and a third in 10% corn sugar. All three lots were then smoked in a uniform manner. The color of the surface of the butts soaked in corn sugar was far superior to that of the other two lots. When cut, the improved color was found to extend into the butt about an inch. Analyses confirmed that the sugar had penetrated to that distance. The investigators concluded that reducing sugars play a role in the quality and stability of the color of cured meats.

Reducing Sugars. The red color of cured meat is due to the nitric oxide–myoglobin formed by the action of nitric oxide on the myoglobin pigment of meat. Haldane (12) and Brooks (2) have shown that reducing conditions are necessary for the formation of nitric oxide hemoglobin from nitrite and hemoglobin. Neill (23), Brooks (3), and Hewitt (13) have shown that dextrose will establish reducing conditions in different biological systems. Jensen (15) pointed out that reducing conditions are essential for proper color development in cured meats.

Following the observations of the effect of reducing sugars on the color of pork butts, Greenwood, Lewis, Urbain, and Jensen (9) investigated the manner in which the reducing sugars affect the color of cured meats. Oxyhemoglobin solutions were diluted with curing ingredients and distilled water and stored in stoppered flasks or test tubes at 5° C. The solutions were analyzed and color observations made at intervals throughout a period of time comparable to meat curing practice. When the sugars—sucrose, maltose, lactose, dextrose, levulose, mannose, galactose, or xylose—were added to dilute oxyhemoglobin solutions, the color changed from red to purple more quickly than when no sugar was added. The color of solutions containing dextrose, levulose, mannose, and galactose changed at the same time, while those containing sucrose, maltose, lactose, and xylose required a longer time to change color. When tetramethylglucose was added, no effect was noted.

Microorganisms. Tetramethylglucose is not utilized by microorganisms. That the effect of the sugars is dependent on their utilization by microorganisms and not on the presence of the sugars per se was demonstrated by additional experiments. When toluene, which inhibits the growth of the organisms, was added to the tubes in a similar series, the color did not change from red to purple. Similar results were obtained with another series in which sterile solutions were used.

Brooks (2) showed that the rate of reaction of nitric oxide with hemoglobin to form nitric oxide hemoglobin was greatest at pH 5.15 to 6.4. An acid condition and a reducing condition are necessary for good color formation in cured meats. It seems well established that the action of microorganisms on sugar, particularly readily fermentable sugar, results in a rapid development of reducing conditions favorable to good color formation in the meat. The exact mechanism occurring in the complex meat curing pickle is not known.

It has been shown that hydrolysis of sucrose to reducing sugars begins about 10 days after the start of the cure (8, 9). Because dextrose is more readily fermented than sucrose by many organisms, there may be an advantage in using some dextrose in the cure. This may help to ensure good reducing conditions at the start of the cure. It would seem to be of most importance in the short cures and in the nitrite and mixed cures. The amount of dextrose used in curing meat has increased in recent years.

Effect of Sugar on Putrefaction

Sugars play an important role in controlling putrefaction during the curing process. This is of most importance in the longer cures and where the cover pickle is re-used. Greenwood and Griffin (8) found that the pH of pickle containing no sugar and used for curing hams rose rapidly after 34 days. The pickles containing dextrose and sucrose remained acid throughout the cure. Some of the hams cured without sugar had a putrid odor, while those cured with sugar were free from such odor.

Greenwood and Stritar (10) studied the effect of varying the amount of sugar in the pump and cover pickle on the re-use of the pickle in a 9-day cure. When the cover pickle was used only once, the presence of sugar in either the cover or pump pickle, together or separately, had no appreciable effect on the quality of the hams. When the cover pickle was reclaimed and re-used, the presence of sugar prevented the growth of putrefactive organisms. The presence of sugar in the cover pickle also had a sparing action on the nitrate and nitrite. Sugars seem to play an important part in maintaining an acid reaction and in controlling the growth of putrefactive organisms, presumably by favoring the growth of fermentative types of bacteria. This is of more importance in the longer cures. Sucrose or dextrose can fulfill this function under practical conditions.

Attempts to replace sucrose with dextrose in the curing of bacon have been unsuccessful. The quality of bacon cured with dextrose is entirely satisfactory, except that it turns brown too readily in frying. Some lots of bacon fry brown when only sucrose and no reducing sugar is used in the cure. Sucrose is hydrolyzed by weak acids, yeasts, and some bacteria. If sufficient sucrose is hydrolyzed, the resulting reducing sugars will cause the bacon to fry brown. Where only cane sugar was used in the curing of pork butts, hams, and bacon, Lewis (18) found the amount of invert sugar expressed in per cent of the total sugar contained in the cured product varied from 2% to as high as 59%. He found that the tendency of bacon to fry dark was related to the amount of reducing sugar contained in the bacon.

The variations in color during frying apparently are due chiefly to the degree to which sucrose is split to reducing sugars during the curing process. Bacon cured with no sugar contained an apparent 0.067% of reducing sugar, with cane sugar 0.159%, and with different mixtures of cane and corn sugar from 0.204 to 0.656%. When fried under uniform conditions, the order of increasing color was identical with the order of increasing reducing sugar content. The time required to reach the same brown color when fried at 275° F. decreased from 7 minutes to less than 3 minutes as the reducing sugars increased from less than 0.10 to 0.65%. The temperature required to reach the same brown color when fried for 2.5 minutes decreased from 345° to 285° F. as the reducing sugars increased from less than 0.10 to 0.65%. Lewis, Oesting, and Beach (19) added different amounts of corn sugar, invert sugar, and cane sugar to ham patties made from hams cured without sugar, and noted their tendency to brown when fried. No effect was noted at concentrations below 0.12%. At concentrations above 0.12% either corn sugar or invert sugar (both reducing sugars) produced a browning effect, while cane sugar did not.

Spray-dried hydrolyzed starch sirups made from potato or cornstarch have been used in Germany to replace cane or beet sugar (7). They were first used in fresh sausage in 1950 and later in cured products. Gisske (5) found that where dried starch sirup was used in sausage, the moisture loss in processing was 8 to 10% less. Optimum amounts were between 0.2 and 0.3%. Larger amounts affected the flavor adversely.

Binders

About 35,000,000 pounds of flours and starches are used annually as binders in meat processing (6), chiefly to absorb moisture resulting from the ice used in chopping. A satisfactory binder must hold this moisture throughout processing (chopping, curing, smoking, cooking, and chilling) and storage of the product. Binders also serve as emulsifying agents between the protein, fat, and moisture of the meat.

The binder must not impart any undesirable flavor or odor to the product. Two types of carbohydrate binders are used, the pure starches and the flours.

Potato starch is regarded as one of the best binders for bologna and frankfurters. Although its primary water absorption is not especially high, it has a low gelatinization temperature, which permits it to absorb water in the early stages of the cooking and smoking operations. Tapioca flour has water-absorbing properties similar to potato starch. The presence of protein in the cereal flours contributes to the binding properties. In this respect corn and wheat flours are superior to their respective starches. Corn, wheat, and rice starches have been used as binders, but they do not perform as well as potato starch. They have poor primary absorption and high gelatinization temperatures. Wheat flours made from ordinary hard spring or winter varieties have not proved very satisfactory as sausage binders. The first clear fraction from milling durum wheat is used extensively as a sausage binder. It has high primary absorption and good moisture-retaining properties. Gelatinized dried flours have high primary absorption but low secondary absorption. They will take up water well during chopping, but lose it rapidly during smoking and cooking.

In addition to the carbohydrate type of binders, soybean flour and dried milk are used in meat processing. The Federal Meat Inspection Division limits the amount of cereal or binder that may be added to sausage to 3.5%. Natural gums, such as Irish moss and karaya, are used in quantities of less than 1% as an emulsifying agent.

Ante-Mortem Feeding of Sugar

Recently the effect of ante-mortem feeding of sugar on the quality of meat has been studied by a number of investigators. Madsen *(22)* in Denmark found that, when pigs were fed 1 to 3 kg. of sugar the day before slaughter, the meat and the liver contained larger amounts of glycogen. They also reported that the flavor of the meat was improved.

Gibbons and Rose *(4)* in Canada found that ante-mortem feeding of sugar to pigs increased the weights of the liver and the glycogen reserves in the liver and the muscle tissue. After slaughter, the glycogen of the muscle tissue is broken down to yield lactic acid. The pH of muscle tissue of the animals not fed sugar varied from 6.0 to as high as 6.6, while that of the pigs fed sugar was about 5.3. The color of the cured meat from the pigs fed sugar was better and more stable. The meat was also less subject to spoilage by bacteria. After smoking, the color differences were less apparent. Gibbons and Rose concluded that the quality of the unsmoked Wiltshire sides would be improved greatly by ante-mortem feeding of the pigs.

Wilcox *et al. (26)* obtained an increase in dressing percentage and sugar content of muscle and liver, and improved texture and flavor of the liver when pigs were fed sugar for 3 to 14 days prior to slaughter. When beef cattle were fed sugar 3 to 12 days prior to slaughter, there were a slight increase in dressing weight, and an increase in the weight of the liver, the sugar content of liver and muscle, and keeping time of the cuts of beef. No significant effect was found on the tenderness, flavor, or palatability of pork or beef.

Literature Cited

(1) Brady, D. E., Smith, F. H., Tucker, L. N., and Blumer, T. N., *Food Research,* **14**, 303–9 (1949).
(2) Brooks, J., *J. Proc. Roy. Soc. (London)*, **B123**, 368–82 (1937).
(3) Brooks, M., *Proc. Soc. Exptl. Biol. Med.*, **32**, 63–4 (1934).
(4) Gibbons, N. E., and Rose, Dyson, *Can. J. Research*, **F28**, 438–50 (1950).
(5) Gisske, W., *Fleischwirtschaft*, **2**, 273 (1950).
(6) Glabe, E. F., *Trans. Am. Assoc. Cereal Chemists*, **2**, No. 1, 20–4 (1943).
(7) Grau, Reinhold, *Die Starke*, **3**, 112–15 (1951).
(8) Greenwood, D. A., and Griffin, H. V., Am. Meat Inst., *Proc. Operating Chem. Sections*, **1938**, 20–31.
(9) Greenwood, D. A., Lewis, W. L., Urbain, W. M., and Jensen, L. B., *Food Research*, **5**, 625–35 (1940).

(10) Greenwood, D. A., and Stritar, J. E., Dept. Sci. Research, Am. Meat Inst., *Bull.*
     SR-30 (1944).
(11) Gross, C. E., *Ibid.,* **1939**, 9–15.
(12) Haldane, J. B., *J. Hyg.,* **1**, 115–22 (1901).
(13) Hewitt, L. F., "Oxidation-Reduction Potentials in Bacteriology and Biochem-
     istry," 4th ed., pp. 52–74, London County Council, England, 1936.
(14) Hoagland, Ralph, U. S. Dept. Agr., *Bull.* **928** (1921).
(15) Jensen, L. B., U. S. Patent 2,002,146 (1935).
(16) Lewis, W. L., Am. Meat Inst., *Proc. Operating Chem. Sections,* **1936**, 7–16.
(17) *Ibid.,* **1937**, 60–7.
(18) Lewis, W. L., Dept. Sci. Research, Am. Meat Inst., *Bull.* **SR-20** (1939).
(19) Lewis, W. L., Oesting, R. B., and Beach, G. W., "Preliminary Report on Place
     of Sugar in Curing Meat," Dept. Sci. Research, Am. Meat Inst., 1936.
(20) Lewis, W. L., and Palmer, C. S., Dept. Sci. Research, Am. Meat Inst., 1925;
     "Meat Through the Microscope," pp. 232–8, Institute of Meat Packing,
     University of Chicago, 1940.
(21) Mighton, C. J., Am. Meat Inst., *Proc. Operating Chem. Sections,* **1936**, 22–8.
(22) Madsen, Jens, *Nord. Jordbrugsforskng.,* **1943**, 340–6; *Chem. Zentr.,* **1**, 1339
     (1944).
(23) Neill, J., *J. Exptl. Biol. Med.,* **41**, 535–49 (1925).
(24) Oesting, R. B., Beach, G. W., and Lewis, W. L., "Study of Curing Pork Butts
     with Cane and Corn Sugar," Dept. Sci. Research, Am. Meat Inst., 1935.
(25) Weiss, F. J., "Use of Sugar in Meat Curing, A Statistical Survey," Sugar Re-
     search Foundation, Inc., 1948.
(26) Wilcox, E. B., Greenwood, D. H., Galloway, L. S., Merkley, M. B., Binns, W.,
     Bennett, J. A., and Harris, L. E., *J. Animal Sci.,* **12**, 24–32 (1953).

RECEIVED March 9, 1953. Journal paper 70, American Meat Institute Foundation.

Sugar in Frozen Foods

WILLIAM F. TALBURT

Western Regional Research Laboratory, Albany, Calif.

Substantial changes in the use of sugars by the frozen food industry have occurred during the past 30 years. The most noticeable trends have been toward the use of lower percentages of sugars, made possible through the use of more rapid freezing schedules, and toward the use of a greater variety of sweetening agents. In 1950 the frozen food industry used approximately 94,000,000 pounds of sweetening agents.

The frozen food industry is one of the youngest and most rapidly growing in the field of food processing. While relatively small as compared with the canning industry, it is becoming increasingly important. In 1942 the volume of the frozen fruit pack was about 5% of the canned fruit pack; in 1948 it was 14% (9). The comparable increase in frozen vegetables during the same period was from 3 to approximately 13%. Production of frozen fruits for the years 1942 to 1951 are shown in Table I. These figures do not include frozen concentrated fruit juices.

Any discussion of the use of sugar in frozen foods should include a little of the early history of this rapidly growing industry as well as statistics on the amount and types of products packed. Discussion here is limited to frozen fruits, as frozen vegetables are not packed with sugar.

The frozen food industry is perhaps somewhat older than is generally known. We usually think of the industry as having been born in the early 1930's. However, this happens to be the time when most of us first became familiar with frozen foods, both fruits and vegetables, in retail-sized packages. For at least five years prior to that time the frozen pack industry had been packing each year in excess of 50,000,000 pounds of fruit, mostly in barrels, for sale to the preserving and ice cream industries. While some confectioners, bakers, and restaurateurs froze small fruits for making ice cream and pastries in the earliest years of the century, the first commercial pack of frozen fruits is generally thought to have been made about 1909 or 1910. Production increased fairly rapidly in view of the storage and transportation difficulties involved and reached 85,000,000 pounds by 1930.

In those early days, sugar played a much more important role as a preservative than it does at present. At that time, fruit was packed mostly in barrels and frozen in rooms maintained at 15° to 20° F. Under those conditions, several days were required to lower the temperature of the fruit to the point where fermentation and mold growth would not occur. Thus it was necessary to use high percentages of sugar, usually 2 parts of fruit to 1 part of sugar (2 + 1) or even 1 + 1, in order to permit freezing without excessive fermentation. This procedure was not always entirely safe, and minor explosions occurred not too infrequently, in which warehouses and refrigerator cars were generously coated with the contents of the barrels. Now that nearly all of the frozen fruit pack is going into containers holding 30 pounds or less and freezing temperatures above −10° F. are very seldom used, such high concentrations of sugar are not necessary. At present very little frozen fruit is packed with more than 20% sugar (4 parts of fruit to 1 part of dry sugar) or 25% sugar sirup (3 parts of fruit to 1 part of sugar sirup). Apple slices and blueberries are usually frozen without any added sugar.

To obtain some idea of the amount of sugar that is consumed by the frozen food industry, it may be well to look at statistics of frozen foods for several recent years.

No data have been found on the amounts of various sugars used by this industry. Data compiled by the Production and Marketing Administration, U. S. Department of Agriculture, covering the industrial usage of sugar, report consumption by the canning, bottling, frozen food, and preserving industries under one heading without a breakdown by individual industries. If one considers the pack of frozen fruit for 1950 (Table I) amounting to approximately 472,000,000 pounds (not including concentrated juices) and assumes that the total pack was 20% sugar by weight, one arrives at about 94,000,000 pounds, or 47,000 tons of sugars used by the frozen food industry during that year. While this figure is not exact, it gives a general idea of the amount of sugar required by the industry. During this same year the total consumption of sugar in the United States was considerably in excess of 7,000,000 tons. Thus the frozen food industry actually required less than 1% of the total sugar consumed in the United States.

Table I.   U. S. Pack of Frozen Fruits and Berries[a]
(Not including frozen concentrated juices)

Year	Pounds
1942	194,643,105
1943	187,266,859
1944	323,886,354
1945	427,037,945
1946	519,092,956
1947	343,519,846
1948	369,722,641
1949	354,020,662
1950	472,173,104
1951	415,944,546

[a] National Association of Frozen Food Packers.

Other pertinent statistics on the pack of frozen fruits are included in Table II, which shows the distribution of the pack by commodities. Strawberries comprise more than one third of the total, and strawberries and red tart cherries make up considerably more than 50% of the total national production of frozen fruits.

Table II.   Frozen Fruit Pack by Product (1951)[a]

	Pounds
Apples and applesauce	28,771,823
Apricots	9,868,584
Cherries, red tart	99,281,673
Peaches	32,380,150
Blackberries	14,573,987
Blueberries	13,920,999
Boysenberries	9,309,409
Raspberries, black	9,559,151
Raspberries, red	19,413,645
Strawberries	157,728,674
Miscellaneous	21,136,451
Total	415,944,546

[a] National Association of Frozen Food Packers.

Many of the packer's problems related to the use of sugars are similar to those encountered by other food manufacturers. Price differentials between various sweeteners, hygroscopicity, relative sweetness, in-plant handling difficulties, and psychological factors, which are much the same for many food manufacturers, are not discussed here. Problems more or less peculiar to the frozen food industry, such as choice of dry or liquid sugars, optimum ratio of fruit to sweetener, and choice of sweetener or combinations of sweeteners, are discussed at some length. In some respects, these problems are similar to those encountered in the preserving and canning industries and, to a lesser extent, in the ice cream industry. However, differences are sufficiently great to warrant a discussion of the factors that are considered by the frozen food packer in selecting the particular sweetener or sweeteners that he uses.

Choice of Dry or Liquid Sweeteners

Sugars are added to frozen fruits for reasons other than to impart the desired sweetness. Particularly in bulk-pack fruit, which is used almost entirely for remanufacturing purposes, sugar is added primarily for protection, as it would be much cheaper for the user to buy unsugared frozen fruit and add all of the required sugar during remanufacture, were it not for the protection from deteriorative changes obtained through addition of sweetener at time of packing. A somewhat different situation prevails in packing retail or consumer packages of frozen fruit, where sufficient sugar must be added during processing to impart the sweetness desired by the consumer. End use of the retail packages demands that sugar be added to the fruit during processing. Here again sugar or sugar sirup, perhaps with small amounts of antioxidants, protects the fruit from discoloration and other deteriorative changes while on its way to the ultimate consumer.

Results of numerous investigations in this field, in combination with the many years of experience of commercial packers, have led to rather standardized procedures in the packing of frozen fruits. Such was not the case during the 1920's and early 1930's, when considerable differences of opinion existed regarding the use of liquid or dry sugar in frozen packs. Joslyn *(4)* mentions the following advantages of sirup over dry sugar: Discoloration due to oxidation is minimized; sirup is more convenient to handle; there is less damage to fruit during packing; there is little or no change in fruit volume and no settling; sirup is a better aid to preservation during freezing than sugar, and can be chilled before use to act as a precooling agent; texture of the thawed fruit is better; and sirup pack is applicable to all fruits.

Despite the excellent reasons for use of sirup as the sweetener in frozen fruits, and the fact that most processors may prefer to use sirup where the end use permits, apparently the disadvantages outweigh the advantages. At the present time 80 to 90% of the frozen fruit is packed with dry sugars. This includes practically all of the fruit in large containers (30 pounds and over) and the smaller containers of frozen strawberries. Most of the other retail packages of fruit, including peaches and raspberries, are sirup-packed.

In a survey reported by the U. S. Production and Marketing Administration in 1951 *(3)* frozen food processors indicated that one of the principal advantages of dry sweetened packs is their general adaptability to any sort of commercial or household end use. For this reason, many smaller firms specializing in only one type of pack generally prefer to use dry sweeteners.

Proposed definitions and standards of identity and standards of fill of containers for frozen fruits *(10)* permit the packing of fruits included in the standards either in sirup or with dry sugars. Composition of sirup and dry sugars must conform to several definite requirements specified in the standards.

Ratio of Fruit to Sweetener

The ratio of fruit to sweetener in frozen foods varies rather widely with the individual fruit, type of sugar used (dry or liquid), end use of the fruit, and individual preference of the customer. Retail packages of frozen fruits are generally packed with 4 parts of fruit to 1 part of dry sugar or 3 parts of fruit to 1 part of sirup. Density and composition of the sirup may be varied within limits for various fruits to give the desired sweetness. No such general statement is possible with bulk-packed fruits, as they are packed to a large extent to the specification of customers. However, a high percentage of this fruit is packed in dry sugar, usually with not more than 1 part of sugar to 4 parts of fruit.

Proposed standards of identity *(10)*, which include all of the common frozen fruits except apples, specify definitely the ratios of fruit to sweeteners to be used and the labeling which should be used to indicate type, amount, and density of packing medium.

Choice of Sweetening Agents

The packer of frozen fruits has considerable latitude in choosing a sweetening agent or combination of sweetening agents. That sucrose is the most widely used of all the sweetening ingredients employed in the packing of frozen foods was found to be the case in a survey conducted in 1948 by the U. S. Production and Marketing Administration (3) covering 31 companies that process 50 to 60% of the national pack of frozen foods.   Of the 31 companies 28 reported that they used only sucrose in their operations, 2 used sucrose and dextrose, and 1 used sucrose and high-conversion corn sirup.   Since this survey was made, there may have been a slight shift toward the use of other sweeteners by some of the packers.

During the past 10 years the advantages and disadvantages of various sweeteners and combinations of sweeteners in frozen foods have been studied.   Joslyn *et al.* (5) at the University of California made several rather detailed investigations of the use of sugars in the packing of frozen fruits.   These studies included not only effects of various sugars on the color, flavor, texture, and drained weight of fruits during frozen storage and subsequent thawing, but, in addition, the effects of these sugars on rate of oxidation of aqueous solutions of ascorbic acid. This latter factor is important to packers, as ascorbic acid added to several fruits during processing to retard enzymatic browning constitutes an appreciable part of the cost of these frozen fruits.   To reduce the amount of ascorbic acid added, it is advantageous to minimize its loss during storage.   Joslyn *et al.* (5) found that 40° Brix dextrose sirup is not suitable for apricots, peaches, or nectarines because of excessive crystallization of dextrose hydrate and enzymatic darkening of the fruit.   This was not true in high-conversion corn sirup packs, in which color retention was somewhat better than with sucrcse.   Little inversion of sucrose was found during frozen storage.   On the basis of taste panel appraisal, the color, flavor, and texture of frozen apricots and peaches were reported to be influenced by degree of inversion of sucrose.

In general, liquid sugar containing more than 50% invert sugar was unsatisfactory for the fruits tested.   However, proposed standards of identity for frozen fruits (10) state that invert sugar is odorless and flavorless, and may be used, inverted or partly inverted, alone or together with sugar in a liquid packing medium for frozen fruits.   In another investigation covering ascorbic acid retention, Miller and Joslyn (6) reported that the degree of protection to ascorbic acid does not vary greatly with type of sugar used.   Conditions used were not intended to simulate those found in frozen foods, as rates of oxidation were determined at room temperature, but they may indicate conditions to be found in frozen fruits.   Under these same conditions evidence was obtained that abnormally high amounts of copper and iron in sweeteners accelerate the oxidation of ascorbic acid.

Somewhat similar results were obtained by Caul *et al.* (2) in a study of the effects of replacing sucrose in frozen peaches by other sweeteners.   Enzyme-converted corn sirup (62 D.E.), corn sirup unmixed (42 D.E.), invert sirup (50% invert), dextrose, and sucrose were used in the tests.   No crystallization of dextrose hydrate was found in 45° Brix liquid packing medium containing 45% detrose and 55% sucrose nor in 55° Brix liquid packing medium containing 25% dextrose and 75% sucrose.   Panel differentiations were made primarily on a basis of differences in sweetness and sourness rather than on flavor characteristic of the sweeteners themselves.

Considerable work along this line has been under way for the past few years at the Oregon Agricultural Experiment Station.   Investigations on the use of dextrose to replace sucrose in frozen foods led these investigators to the conclusion that in 50° Brix sirup not more than 20% of the sucrose solids should be replaced with dextrose because of low solubility.   However, Sather and Wiegand (8) reported that replacement of 40 to 50% of sucrose solids with corn sirup (90 to 95% regular conversion corn sirup and 5 to 10% refiner's sirup) in a 40° to 50° Brix sirup resulted in a significantly higher drained weight and a frozen fruit superior in texture, flavor, and color.

Another interesting problem related to the composition of packing medium in frozen fruits has been reported by Rabak and Diehl (7).   Occasionally there forms on frozen fruits packed in sucrose a crystalline deposit which, in many cases, resembles a heavy growth of mold.   While this condition has not been so prevalent as to cause widespread consumer complaints, it has been reported by several packers.   This "moldlike" crystalline growth has been identified by Young and

Jones *(11)* as sucrose hemiheptahydrate. They also made phase studies of the sucrose-water system at low temperatures and found evidence of several additional hydrates of sucrose. As a result of this work, a public service patent was obtained *(12)* covering the glazing of fruits with sucrose polyhydrates in order to protect the fruit during storage and handling. Procedures have been developed by Brekke and Talburt *(1)* for preventing sucrose hydrate formation in certain frozen fruit products through replacing 20 to 30% of sucrose solids with invert sirup.

One problem that is currently being investigated at the Western Regional Research Laboratory, though not primarily designed to study sugars in frozen foods, may result in some interesting developments in this field. Comprehensive storage on frozen foods under conditions simulating those found during storage and distribution of these items are under way. These studies, which will extend over a period of several years, are designed to furnish data on the maximum permissible times and temperatures that can be safely used in the warehousing and distribution of frozen foods without causing undue quality deterioration. In this work commercial samples of frozen foods are obtained from all the main producing areas in the United States, prepared from the leading commercial varieties in the areas. As commercial samples are used throughout the work, samples taken from different plants and even from the same plants will vary in composition of the packing medium and this factor will be one of the important variables in the experiments.

Immersion Freezing

Although only a very small amount of fruit is frozen by immersion, this procedure may be of interest to those in the sugar industry. In this type of freezing, small fruits, such as strawberries or cane berries, are washed, graded, and immersed, for the short period required for freezing, in a refrigerated liquid maintained at about 0° to 10° F. The immersing liquid usually is a rather concentrated sugar solution. The fruit is frozen rapidly and has a thin coating of ice which is reasonably effective in protecting from oxidation during storage. For certain uses immersion-frozen fruit is preferable to conventionally frozen fruit and has in the past sold for a slightly higher price. Although several types of immersion freezers have been built and used on a limited scale, technical and operational difficulties have apparently prevented their widespread adoption. One problem with certain types of immersion freezers is the progressive dilution of the immersing medium resulting from extraction of liquid from the fruit or from water adhering to the surface of the freshly washed fruit.

Proposed Standards of Identity and Standards of Fill

The proposed standards of identity for frozen fruits and fill of container, issued by the Food and Drug Administration and published in 1950 *(10)*, are an important milestone in the development of the industry and are a definite sign that the frozen food industry is growing up. Some of the proposed standards of identity may be changed in their final form. In the hearings that preceded the issuance of these proposed standards, approximately 5000 pages of testimony were taken; about 1000 pages related to the use of sugar in frozen foods. Transcripts of these hearings are available through a private reporting company to interested persons.

Summary

As a result of years of research on the use of sugars in frozen foods, and the experience of frozen-food packers in using various types of sweeteners, several generalizations can be made.

Where only one type of sweetening ingredient is to be used in a plant, sucrose as a rule is the most acceptable, as sucrose does not limit the end use of the product, whether for remanufacture or for retail trade. This is the deciding factor on choice of sweetener in many small plants, where volume of production is not sufficient to warrant added cost of handling two types of sugars.

Dextrose, dextrose hydrate, or corn sirup (42 D.E. or above) can be satisfactorily used to replace up to one third of sucrose in most of the frozen fruits commonly packed without adversely affecting color, flavor, or texture of product.

Where end use permits, high-conversion corn sirup or corn sirup solids, alone or in combination with other sweeteners, can be satisfactorily used in the packing of frozen fruits.

Literature Cited

(1) Brekke, J. E., and Talburt, W. F., *Food Technol.*, **4**, 383–6 (1950).
(2) Caul, J. F., Sjöström, L. B., and Turner, W. P., Jr., *Quick Frozen Foods*, **13** (4), 54–8 (1950).
(3) Jones, P. E., and Thomason, F. G., U. S. Dept. Agr., *Inform. Bull.* **48**, 160 (June 1951).
(4) Joslyn, M. A., *Food Inds.*, **2**, 350–2 (1930).
(5) Joslyn, M. A., Jones, W. L., Lambert, E., Miller, J., and Shaw, R. L., *Quick Frozen Foods*, **12**, 72–4, 124–6 (March 1950).
(6) Miller, J., and Joslyn, M. A., *Food Research*, **14**, 340–53 (1949).
(7) Rabak, W., and Diehl, H. C., *Western Canner and Packer*, **36** (4), 55 (April 1944).
(8) Sather, L., and Wiegand, E. H., *Quick Frozen Foods*, **10**, 81–3, 107–8 (May 1948).
(9) U. S. Dept. of Commerce, "Appraisal of the Competitive Position of Fruits and Vegetables," p. 45, July 1949.
(10) U. S. Food and Drug Administration, *Federal Register*, **15**, 6674–86 (Oct. 4, 1950).
(11) Young, F. E., and Jones, F. T., *Phys. Colloid Chem.*, **53**, 1334–50 (1949).
(12) Young, F. E., and Jones, F. T., U. S. Patent 2,542,068 (Feb. 20, 1951).

RECEIVED March 30, 1953.

Effects of Carbohydrates and Other Factors On Color Loss in Strawberry Products

E. EVERETT MESCHTER

The American Preserve Co., Philadelphia, Pa.

Pigment extracted from strawberry juice with 1-butanol was used to study effects of various materials on the loss of color. The data show that temperature and pH value have a great effect on the rate of pigment loss. Salt concentration affects the rate of pigment loss in strawberry juice. The effect seems to be related to the presence of sugars, as a similar effect is not observed in the reaction with extracted pigment when no sugars are present. Both ascorbic acid and dehydroascorbic acid change the magnitude of the rate of pigment loss. Both furfural and hydroxymethylfurfural increase the rate of pigment loss, and because these compounds are typical of sugar degradation products, it is suggested that sugar deterioration products react with the pigment of strawberry preserves. Materials present in strawberry juice itself (probably sugars) apparently have an appreciable effect on rate of pigment loss.

Strawberry preserves represent about 26% of all fruit preserve flavors, or about 12% of the total production of all fruit spreads, and consequently the largest volume of a single variety of fruit spread in the country.

When strawberry preserves are made by modern vacuum pan methods, the finished product is a bright red or bright maroon-red, depending on the variety of strawberry used. As the preserve is stored at room temperature in a warehouse or on a grocery shelf, the pleasing red color diminishes, so that within six months the product has lost so much of its appealing color as to become deep maroon-brown and be unsalable. The author's approach to this problem has been based on measurement of kinetics of the pigment system in various environments in the presence of different sweetening agents.

The first step in this investigation was to isolate the pigment from strawberry juice by the method described by Sondheimer and Kertesz (11). This method involved the saturation of the juice with salt and extraction with 1-butanol, then concentration of the butanol extract under vacuum and in a nitrogen atmosphere. The anthocyanin concentrate was then taken up with hydrochloric acid in anhydrous methanol, precipitated with ether, dissolved in 0.01% hydrochloric acid, and saturated with picric acid. The anthocyanin picrate crystallized out of this solution upon storage at 0° C. as lustrous reddish bronze prisms which were further purified by recrystallization.

The purified anthocyanin chloride, believed to be pelargonidin 3-monoglucoside, obtained by this procedure, had external characteristics which were identical with the product prepared by Sondheimer and Kertesz.

However, the chromatogram on paper of this purified material (by ascending technique using 4 parts of 1-butanol, 1 part of glacial acetic acid, and 5 parts of water as the solvent) showed only one red-orange spot, while the filtered, water-soluble portion of the pigment just prior to the picration step showed not only the same red-orange spot but also a purple spot which is perhaps an isomer of pelargonidin 3-monoglucoside (2) just below and not imposed upon the larger, more intense red-orange pelargonidin spot. (The study of the nature of this purple spot has not been completed.) It seemed evident, therefore, that the purification by crystallization as the picrate had left behind a portion of the pigment normal to the original juice.

For this reason the basic coloring material, wherever extracted pigment was used, was the water-soluble portion of the ether-precipitated concentrate of the 1-butanol extract of strawberry juice (St B). This material has not been further purified by picration and crystallization. It contains essentially all the anthocyanin coloring pigments in strawberries free from the natural sugars. It was found, however, that the reactions of the water-soluble portion and the recrystallized material were not significantly different.

The natural strawberry material was prepared from a good commercial grade of frozen Blakemore strawberry juice. Its history showed that it had never been pasteurized, was pressed on stainless steel equipment, and had not been otherwise abused. The frozen juice was slowly thawed and the early free-run juice was collected at a soluble solids content of 22% (St J).

This made it possible to conduct experiments allowing more than 2 to 1 dilution for incorporation of other reacting material and buffer solutions, and still maintain the strawberry ingredients in concentration normal to the original natural juice.

The color of some of this concentrated strawberry juice was destroyed by autoclaving at 121° C. (250° F.) for 1 hour. This material (St JD) was brown colored, slightly turbid, and identical with the concentrate (St J), except for the damage done by the heat treatment.

Some of the concentrate was diluted back to single-strength strawberry juice and saturated at room temperature with sucrose. This mixture (St S) closely simulates the composition of a finished strawberry preserve.

In all cases, the storage temperature for this study was 38° C. (100° F.) and the pH value was maintained at 3.0 ± 0.03. All juice and buffer solutions stored at 38° C. were saturated with thymol for prevention of fermentation and mold growth. Thymol has been found to be clearly inert as far as the color loss is concerned. Buffer solution in all cases refers to Sorensen's citrate buffer, adjusted to pH 3.0 with hydrochloric acid, and saturated with thymol. The importance of buffer concentration is shown below.

The nitrogen content of all the materials added to the strawberry juice was reported by a consulting analytical laboratory to contain a trace or no nitrogen as determined by the Dumas method.

The ascorbic acid content of the frozen concentrated juice was very low—less than 1 mg. per 100 grams as determined by potentiometric titration. Normal ascorbic acid content of fresh strawberries is about 50 to 100 mg. per 100 grams (3). Ascorbic acid may have been lost during the freezing and thawing operations involved in the concentration procedure.

The copper content of the natural juice was found to be about 1 p.p.m. This is approximately the normal amount in natural strawberries and indicates trace or no contamination of the juice by copper in processing.

The raw material, then, consisted of:

1. St B. Water-soluble portion of the ether-precipitated butanol-extracted pigment of strawberry juice.
2. St J. A representative sample of commercial strawberry juice, concentrated by freezing.
3. St JD. The same juice as in (2) damaged by high temperature processing.
4. St S. Single-strength strawberry juice made up to 67% solids with sucrose.

The changes in anthocyanin concentration were followed by a modification of the method of Sondheimer and Kertesz (9), which is based on the measurement of absorbance (optical density) at 500 mμ of solutions of the anthocyanin at pH 3.4 and 2.0. The anthocyanin concentration is proportional to the absorbance difference. Figure 1 shows the spectra of solutions of anthocyanin pigment of the same concentration at different pH values.

The authors have found, however, that at pH 1.0 the absorbance reaches a maximum, and a pH value of less than 1.0 does not further increase the absorbance measurably at its 500 mμ peak. At 5.4 or at any pH values between 5.0 and 6.0, the absorbance of anthocyanin at 500 mμ is at an indistinguishable minimum. Therefore, the critical control of pH is eliminated, and test papers can be used to determine the broad range of pH in which absorbance difference accurately reflects anthocyanin content. The factor for conversion for absorbance difference, pH 2.0 to 3.4 to pH 1.0 to 5.4, is 1.7. This method is believed to be more accurate, less

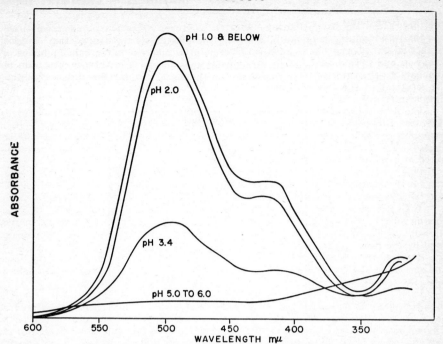

Figure 1.   Change in Spectra of Buffered Solution of Strawberry Juice Pigment Concentrate with Change in pH at Constant Concentration

tedious, and considerably faster than the original method, as it eliminates the need for precise pH measurement.

Anthocyanin content is not the sole factor in determining the acceptability of a strawberry preserve.   Mackinney has pointed out in reports to the National Preservers Association that the pigment content may drop to as low as 10% and still be acceptable, as long as browning development is low and brightness remains high. Even if pigment content is high, its value may be overshadowed by a high browning rate.   Browning increases in rate four times as fast as pigment loss with increasing temperature.   Therefore a preserve stored at 10° C. will be much more acceptable when half of its pigment has been lost than one of the same pigment content that has been stored at 20° C., because browning will be much lower in the lower temperature sample.   Anthocyanin content alone has been used as a measure of color or pigment stability in this paper, even though some samples of higher pigment content were less acceptable visually than others, as the dark brown color prevented realization of the high pigment content.   A method of color measurement for evaluating visual acceptability may have to include a measure of brightness, turbidity, and dominant wave length rather than pigment content alone.

Determination of color change in strawberry products is not a clear-cut, ideally accurate process.   In all measurements of color loss in solutions using only extracted pigment as color source, reaction rates were clear cut and reproducible.   Where natural sugars and complex materials found in all natural products were present, it was necessary to rely on trends and orders of magnitude to evaluate importance of additives in changing reaction rates. For this reason, changes of less than 5% in magnitude are in general insufficient evidence of important effect and are not reported here.

In considering the possible means by which strawberry color is deteriorated, investigation reveals five significant factors:

1.   Effect of temperature
2.   Effect of pH value
3.   Effect of salt concentration
4.   Effect of ascorbic and dehydroascorbic acid
5.   Effect of sugars and sugar destruction products

Effect of Temperature

The most important factor in changing the kinetics of the degradation of color in strawberry products is temperature. The preserver can alter this factor to a limited degree in his choice of manufacturing and storage procedures. The rate of color deterioration increases in proportion to the log of the temperature. Figure 2 shows the relationship graphically.

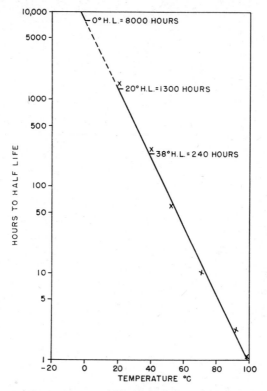

Figure 2. Effect of Temperature on Half Life of Color in Strawberry Preserves

At room temperature (20° C.), the half life of the color of strawberry preserves is about 1300 hours. In the summertime, the grocery shelf in the front of the corner grocery store may reach 38° C., at which temperature the half life drops to 240 hours. It is possible to increase the life of the color in the finished preserves six or seven times over room temperature storage by storing under egg storage conditions (4° C.), where half life is extended to 6000 to 8000 hours (250 to 320 days).

Effect of pH Value

The rate of anthocyanin degradation is greatly affected by pH value. Increasing acidity has a protective effect on the stability of the pigment. Figure 3 shows the relative rate of pigment loss in buffered solutions at various pH values.

It is obvious that pH must be accurately controlled when effects of added ingredients are to be studied   Sorensen's citrate buffer solution was used throughout.

Effect of Buffer Salt Concentration

The importance of maintaining constant buffer salt concentration is graphically illustrated in Figure 4, which shows an increase in rate of pigment loss when increments of sodium citrate were added to strawberry juice (St J). However, this detrimental effect was not observed when sodium citrate was added to extracted pig-

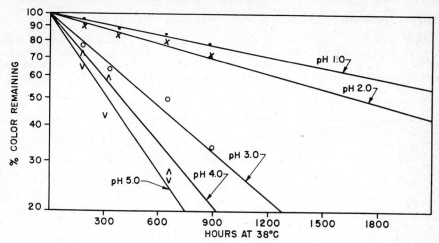

Figure 3. Effect of Change in pH on Rate of Color Loss in Buffered Solutions of Strawberry Juice Concentrate

ment (St B) solutions, as shown in Figure 5. A protective effect of the citrate ion on color loss in the iron-catalyzed anthocyanin–hydrogen peroxide reaction has been shown by Sondheimer and Kertesz (10). A similar mechanism may account for the increased stability of the color with increments of sodium citrate in the sugar-free color extract. In the strawberry juice, however, any slight protective effect shown by the citrate ion is overpowered by the effect of the sodium ion on sugar destruction observed by Moyer (6, 7). The products formed by this degradation may thus be more reactive with the pigment.

Effect of Ascorbic and Dehydroascorbic Acids

The effect of ascorbic acid on color loss in strawberry products has been studied by Sondheimer and Kertesz (10). This reaction is catalyzed by the presence of iron or copper. Although they make it clear that the presence of hydrogen peroxide in strawberry products has not been demonstrated, Sondheimer and Kertesz have suggested that the hydrogen peroxide formed when ascorbic acid is oxidized to dehydroascorbic acid may play a role in the degradation of pelargonidin 3-monoglucoside. Indeed, this may well account for some of the kinetic difference in the reaction between ascorbic acid and the pigment and dehydroascorbic acid and the

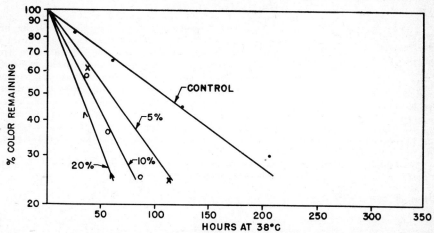

Figure 4. Effect of Sodium Citrate on Rate of Color Loss in Frozen and Thawed Strawberry Juice

pigment. However, the relatively high velocity of pigment destruction with de-hydroascorbic acid suggests that a much more complex mechanism is involved.

The dehydroascorbic acid used in this experiment was prepared by the Research Laboratories of the Hoffmann-La Roche Co., Nutley, N. J., by the process of Pecherer *(8)* and is reported to be 95 to 100% pure. The impurities may include some as-corbic acid and inert oxidation products which result during processing. Even if the maximum impurities consisted entirely of ascorbic acid, the importance of the effect of dehydroascorbic acid on the pigment system would not be significantly changed.

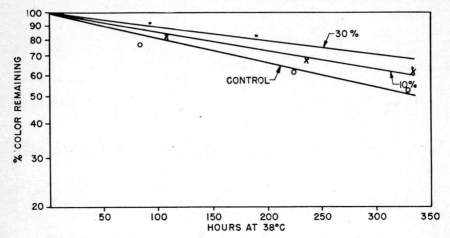

Figure 5. Effect of Sodium Citrate on Rate of Color Loss in Buffered Solutions of Strawberry Juice

The relationship between ascorbic acid and dehydroascorbic acid on the rate of strawberry pigment loss in water-soluble pigment (St B) is given graphically in Figure 6.

The susceptibility of ascorbic acid to oxidation has long been known, but it is surprising to note its stability when strawberry preserves are made by vacuum processing methods.

A sample of strawberries contained 39 mg. (360 p.p.m.) of ascorbic acid per 100 grams as determined by potentiometric titration. After the preserves were sterilized and cooled, 82% of the original ascorbic acid content still remained. After the preserve had been stored for 2 weeks at 38° C., 52% of the ascorbic acid and 41% of the original anthocyanin content remained.

It is obvious that the mechanism of ascorbic acid attack on the strawberry pig-ment is of very practical interest, as the preserver can do nothing about the original ascorbic acid content of his raw material. This also emphasizes the importance of the amount of metallic contamination in the strawberry preserve. Iron and copper in themselves are not important contributors to the pigment loss mechanism, except in so far as they hasten the destruction of the ascorbic acid. These destruction products in turn attack the pigment at an increased rate. If only ascorbic acid, and not its degradation products, were responsible for the rapid loss of color, as the ascorbic acid in the pigment medium was destroyed, the rate of loss of remaining pigment would be expected to decrease. Such a change in rate is not significant and apparently the destruction products of ascorbic acid also react with the color.

Table I. Effect of Ascorbic Acid in Presence of Iron and Copper on Rate of Color Loss in Strawberry Juice (St J)

Ascorbic Acid, P.P.M.	Cu or Fe, P.P.M.	Hours to Half Life
	Control	320
None	Cu, 40	310
None	Fe, 40	310
400	None	190
400	Cu, 10	140

Table I shows the relationship between rate of pigment loss in single-strength straw-berry juice and the metallic ion content with and without ascorbic acid.

Although ascorbic acid, particularly in the presence of iron or copper, is ca-pable of increasing the rate of pigment loss, the rate of loss when none of these agents is present is already of a distressing magnitude. Thus some other cause of color loss has yet to be explained.

Effect of Sugars and Sugar Destruction Products

The data presented in this section attempt to show that several sugars and sugar degradation products are capable of reacting with the pigment components of strawberries to increase their rate of loss materially. Thus substances formed during the browning of sugars may cause color loss in strawberry products. It has been demonstrated that when extracted pigment is stored at various pH values,

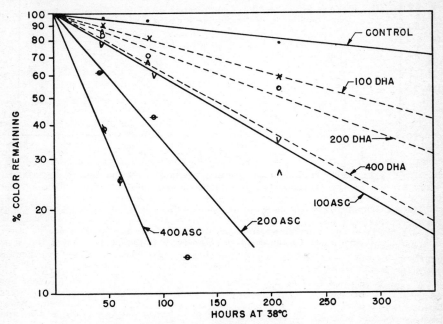

Figure 6.   Effect of Ascorbic and Dehydroascorbic Acids on Rate of Color Loss in Buffered Solutions

Units, p.p.m.

the stability of the color is greatly increased as the pH value is lowered (see Figure 3). When strawberry sirup is similarly stored, however, the stability of the pig-ment reaches a maximum at a pH value of 1.8. Below this value, pigment stability decreases as the pH is lowered. The high rate of browning of sugars at low pH values is well known and the data in Table II suggest that the reactivity of these sugar degradation products with the pigment is more important than the stability contributed to the pigment by the low pH value.

Table II.   Effect of pH Value on Half Life of Color in Strawberry Sirup (St S)

pH	Hours to Half Life
0.30	240
0.65	385
1.05	415
1.55	460
1.80	480
2.10	430
2.55	330
2.88	275
3.18	210

Sondheimer and Kertesz pointed out as early as 1947, in reports to the National Preservers Association, that aldehydes were capable of reacting with pelargonidin

3-monoglucoside. Formaldehyde reacts with the pigment to form a colorless solution and a purple precipitate. Furfural and hydroxymethylfurfural are known to be typical intermediates in the destruction of sugars, and the kinetics of these two products with strawberry color is shown in Figure 7.

These data illustrate the ability of some sugar degradation products to increase the rate of strawberry pigment loss. Furfural and hydroxymethylfurfural are cited only as examples typical of a great number of products which result from the breakdown of sugars.

Further evidence that the products of sugar degradation increase the rate of strawberry pigment loss is presented in Table III. Here are listed, in order of increasing anthocyanin degrading activity, various carbohydrates with their corresponding half lives. This list is essentially the order of ring stability or carbonyl activity demonstrated by these sugars in other reactions of this type *(1, 5)*.

Table III. Relative Effect of 40% Concentration of Various Carbohydrates on Stability of Water-Soluble Pigment (St B)

Carbohydrate	Hours to Half Life
Corn sirup	
Methyl α-D-glucoside	
Mannose	
Glycerol	
Sucrose	650–700
Maltose	
Sorbitol	
Control, no sugar	
Arabinose	
Levulose	300
Sorbose	240

The source of the carbohydrate that might enter into such a reaction need not necessarily be an added one. Indeed, there is evidence that the source material is already present in the strawberry juice itself. Figure 8 shows the effect of the addition of concentrated strawberry juice and heat-damaged strawberry juice upon pigment loss. The juice must contain an agent or agents, which are increased upon

Table IV. Effect of Fermentation upon Rate of Color Loss in Presence of Various Carbohydrates in Strawberry Juice (St J)

Carbohydrate	Hours to Half Life	
	Fermented	Unfermented
Corn sirup	380	250
Sucrose	250	143
Dextrose	250	150
Levulose	168	114
Control, no sugar	205	100

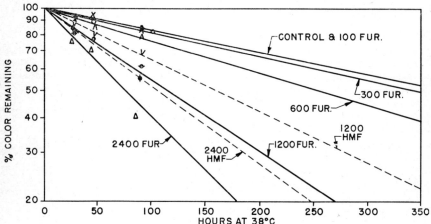

Figure 7. Effect of Furfural and Hydroxymethylfurfural on Rate of Color Loss in Buffered Solutions

Units, p.p.m.

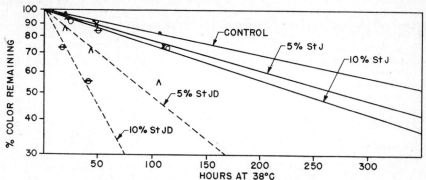

Figure 8.   Effect of Addition of Concentrated Strawberry Juice and Heat-Damaged
Strawberry Juice on Rate of Color Loss in Buffered Solutions

heat treatment, capable of reacting with the pigment. Such a mechanism could also explain the difficulty in making uniform measurement of rate of pigment loss on natural strawberry products, as these unidentified products may be present in varying quantities from time to time.

Further evidence in support of this explanation is given by the data in Table IV.   Here, the rate of anthocyanin loss was measured in regular unfermented strawberry juice in the presence of several sugars.   Then some of the same juice was fermented by *Saccharomyces cerevisiae* and any pigment lost during the fermentation was replaced by the addition of water-soluble pigment.   Carbohydrates were similarly added to this fermented juice after fermentation.   The rate of pigment loss in the two juices was then followed, with the results given.

After most of the natural carbohydrates have been removed by fermentation, the stability of the pigment on storage is greatly increased.   Differences in the reactivity of the various added carbohydrates are evident.   The relative effect of these sugars has always remained in this order, regardless of the environment in which the reactivity of the sugars with the pigment has been studied.

Loss of sugar is not the only change in the juice upon fermentation.   But the data strongly suggest that some ingredient normally reactive with the pigment has been removed and it would be plausible to assume that this ingredient is a carbohydrate.

It is to be expected that under the conditions of these experiments, as well as in finished preserves, some sucrose will be inverted and thus the sucrose samples will contain both dextrose and levulose.

Acknowledgment

The writer is indebted to Gordon Mackinney, University of California, for his vital interest and helpful criticism of the work, and to Sidney M. Cantor, American Sugar Refining Co., for his encouragement and valuable assistance in the evaluation and interpretation of the data.

References

(1) Cantor, S. M., and Peniston, Q., *J. Am. Chem. Soc.*, **62**, 2113 (1940).
(2) Chichester, C. O., University of California, Berkeley, Calif., personal communication.
(3) Hanson, E., and Waldo, G. F., *Food Research*, **9**, 453 (1944).
(4) Kertesz, Z. I., "Pectic Substances," New York, Interscience Publishing Co.
(5) Maillard, L. C., *Compt. rend.*, **154**, 66 (1912).
(6) Moyer, W. W., U. S. Patent 2,270,328 (1942).
(7) *Ibid.*, 2,349,514 (1944).
(8) Pecherer, B., *J. Am. Chem. Soc.*, **73**, 3827 (1951).
(9) Sondheimer, E., and Kertesz, Z. I., *Anal. Chem.*, **20**, 245 (1948).
(10) Sondheimer, E., and Kertesz, Z. I., *Food Research*, **17**, 288–98 (1952).
(11) Sondheimer, E., and Kertesz, Z. I., *J. Am. Chem. Soc.*, **70**, 3476 (1948).

RECEIVED March 17, 1953.

Role of Carbohydrates in Infant Feeding

L. EMMETT HOLT, JR.

New York University College of Medicine, New York 16, N. Y.

With modern methods of infant feeding, infants can be reared successfully on low- and high-carbohydrate formulas. With the exception of a few pathological conditions, one carbohydrate appears to be as good as another, and economy is an important determining factor.

Artificial feeding is a subject of particular interest. From time immemorial carbohydrates have played a leading role in infant feeding. There are two chief eras: the prescientific era, which extended well into the middle of the past century, and what we like to speak of as the modern scientific era. The prescientific era can be dismissed in a few words. From ancient times to the beginning of the modern era, artificial infant feeding was a failure—the great majority of infants died, the mortality varying from 90 to 99%. The need for protein was not appreciated; cow's milk, when given, was grossly contaminated and produced disease, and reliance was placed primarily on fat and carbohydrate. Mixtures of butter and honey were recommended in ancient times. Starchy "paps" made of bread soaked in water or beer were used throughout the middle ages.

The scientific era began in Europe around the middle of the past century. The important advances were the discovery of pathogenic bacteria and means of destroying them, which made cow's milk a safe food; chemical studies of milk, particularly of breast milk, which gave an idea of what the baby received when fed as nature intended; and studies of the nutritive significance of the various food ingredients. Knowledge, unfortunately, came piecemeal rather than all at once. Half truths were discovered and feeding policies based upon them led to a great deal of confusion. A pediatric colleague used to remark that there were three schools of thought in infant feeding: that in which all digestive disturbances were attributed to the protein, that in which they were attributed to the fat, and that in which they were attributed to the carbohydrate. As one of these calorigenic foodstuffs was singled out for opprobrium, the others became relatively glamorous by contrast.

From our point of vantage today we believe we can see many things more clearly than did our predecessors. Many of the older ideas still persist, however, in one form or another.

Carbohydrates in Diarrhea

The chief vice attributed to carbohydrates was their alleged tendency to incite diarrhea. The explanation offered was a very simple one. Fermentative bacteria were known to be present in great numbers in the intestine; these attacked the simple sugars, producing volatile (short-chain) fatty acids. In in vitro experiments on isolated loops of intestine these were reported to stimulate peristalsis. Following this concept of fermentative diarrhea, carbohydrates were to be avoided in diarrhea. The most famous feeding designed for this purpose was Finkelstein's protein milk, which is still used to some extent. Healthy children were given carbohydrate supplements in order to avoid difficulties from protein and fat, but always with misgivings and the fear of a fermentative diarrhea in the background. Sugars that fermented more readily than others were to be avoided. To be preferred were the polysaccharides or partly hydrolyzed polysaccharides which required further hydrolysis and released fermentable sugar so slowly that the bacteria had little chance at it. Mixtures of dextrins and maltose have enjoyed extraordinary popularity among pediatricians.

104

The above concept has had an enormous impact upon medical practice and the industries which serve it. Nevertheless, it is fair to state that as far as a scientific basis is concerned, it is a structure built on sand. The present concept is that infantile diarrhea is caused either by pathogenic bacteria in the intestine or by toxic bacterial products which reach the intestine by way of the circulation—not from the decomposition of normal food. Pediatricians back in the 1920's, who were brave enough to add carbohydrate to the protein milk diet of their diarrheal babies, saw only benefit rather than an exacerbation of the disease. That fermentation of carbohydrate occurred in the intestine is clear enough and short-chain fatty acids were formed, but the present view is that the difficulty in diarrhea is primarily malassimilation rather than indigestion; the fraction which the intestine cannot assimilate falls to the bacteria. Moreover, the short-chain fatty acids formed by bacterial action are not irritating; they can be fed to normal babies without harm. The enthusiasm for malt-dextrin mixtures goes back to the German chemist Liebig, who, when his own child developed diarrhea, tried the effect of feeding dextrinized starch, following which the diarrhea ceased. It was a single observation, but such was the influence of Liebig that this feeding spread rapidly in Germany and, somewhat later, in this country. The malted milk industry as well as the malted infant food industry owes its origin to Liebig.

Nutritional Properties of Sugars

If there are differences in the nutritional properties of different sugars, they are small and applicable only to specific disease situations. Levulose appears to have some advantage over glucose in diabetes mellitus and its superiority as a glycogen former and its higher renal threshold are of advantage in parenteral nutrition. The central role of levulose in sugar utilization is now well known. Certain cells, such as spermatozoa, utilize levulose directly, levulose being the normal sugar of seminal fluid in the human.

Galactose and lactose, though long regarded as physiological for infants, may in certain circumstances prove harmful. A deficiency of the enzyme waldenase, which converts galactose 1-phosphate into glucose 1-phosphate, is found in certain infants, which results in the syndrome of idiopathic galactemia with failure to thrive, an enlarged glycogen-containing liver, and cataract formation. The process is fortunately reversible to a considerable extent, if a galactose-free diet is given.

Carbohydrates in Nutritional Disturbances

A contentious question concerns the effect of various carbohydrates in chronic disturbances of nutrition—notably celiac disease and the various forms of sprue. Clinical observations have suggested to many that polysaccharides may be poorly tolerated in this condition as contrasted with mono- and disaccharides, giving rise to exacerbations of the digestive disturbance. There is some scientific evidence to support this view, although it is far from convincing. A difficulty in deciding such a contentious point is the lack of an accurate method for measuring carbohydrate absorption. Unlike fat and nitrogen, the absorption of which can be accurately determined by balance studies, unabsorbed carbohydrate is broken down and cannot be identified in the feces. The postabsorptive blood sugar curve is an inaccurate measure of absorption, as it is also influenced by excretion, utilization, storage, and glyconeogenesis.

It was felt that tagged carbohydrates could be employed to study some of the unsolved problems of absorption. A balance study of tagged carbon after the administration of carbon-labeled carbohydrate would reveal accurately the extent of absorption. During the past 2 years some studies of this kind have been carried out. They are of a preliminary nature, and all the questions have not yet been answered. The first studies were carried out with carbon-13. Subsequently carbon-14 was released in quantities suitable for studying infants and young children and that isotope is still being used.

Table I illustrates some observations made on infants and young children who were believed to be metabolically normal. The completeness of sugar absorption in these studies was somewhat of a surprise. In contrast to nitrogen, fat, minerals,

Table I.   Absorption of Carbon-14 Sugar by Normal Subjects

Subject	Sugar Given, μc.		% Absorbed
P.G.	Sucrose	20	98.99
P.G.	Glucose	20	99.79
P.G.	Glucose	20	99.96
P.G.	Sucrose	20	99.30
T.L.	Sucrose	20	99.66
T.L.	Sucrose	24.2	99.84
B.A.	Sucrose	5	99.70
B.A.	Sucrose	4.2	98.45

and vitamins, which may on occasion be absorbed somewhat in excess of 90%, sugar seems to be absorbed in excess of 99%.

Some studies were made on premature infants, who are known to have a tendency toward hypoglycemia and who have been suspected of having impaired sugar absorption. Two such observations are shown in Table II. Even in these feeble subjects absorption is extraordinarily complete.

Table II.   Absorption of Carbon-13 Sucrose by Premature Babies

		Baby A	Baby B
Isotopic sucrose given, mg. excess		51.0	120.3
Isotopic carbon in stool, mg. excess		0.83	1.73
Day 1	0.04	(in subsequent	(in subsequent
Day 2	0.36	96 hours)	50.75 hours)
Day 3	0.35		
Day 4	0.06		
Isotopic carbon absorbed, mg. excess		50.17	118.57
% of intake absorbed		98.4 ± 1.6	98.6 ± 1.6

It was of interest to follow the excretion of labeled carbon after the administration of tagged carbohydrate. It was found that only traces of tagged carbon were excreted in the urine, but that the major portion of the isotope appeared in the breath as labeled carbon dioxide. Typical carbon dioxide exhalation curves are shown in Figure 1. The carbon dioxide exhalation curve rises promptly within an hour of ingestion, reaching a maximum and falling to minimal levels within 7 or 8 hours. Thus it would appear that there is a very rapid turnover of carbon in the body after its administration as sugar. The carbon dioxide exhalation curve measures sugar absorption and utilization.

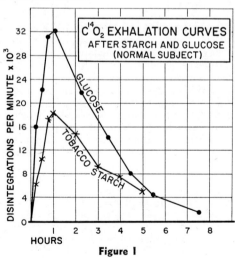

Figure 1

A few comparisons of starch and simple sugars from the point of view of absorption in normals and in two celiac patients indicate a less efficient absorption and subsequent utilization in the case of starch. However, the starch used was prepared from tobacco leaves and may conceivably differ in structure and digestibility from the usual forms employed in human diet. It is planned to carry out such studies with starch prepared from potato. In the meantime the question of the relative digestibility of starch and sugar in the celiac patient must remain unanswered.

Importance of Carbohydrates in Diet

Are carbohydrates essential in human diet? Observations on Eskimo diets of lean meat and fat indicate that health can be maintained on such a diet. Only a single investigator, Chwalibogowski, has attempted to feed infants without carbohydrate. He was able to continue this diet for months, but was forced to discontinue it when an infection and ketosis developed.

The German pediatric literature early in the present century has many references to a condition referred to as *Milchnahrschaden* (milk injury) occurring when infants were fed on cow's milk without added carbohydrate. The clinical picture was an indefinite one—failure to thrive, with pale, constipated stools. It was probably due to presentation of the protein in considerable part in a form which yielded indigestible curds and to a consequent decrease in available calories. Recent observations on formulas without added carbohydrate, but with the milk processed in such a way as to avoid difficulty with curd assimilation, have failed to produce *Milchnahrschaden*.

With modern methods of infant feeding, infants can be reared successfully on low- and high-carbohydrate formulas. There may be minor differences in the effects of such diets, but they are not conspicuous. There may also be differences in the effects of supplementation with different carbohydrates. These are not as yet demonstrated, with the exception of a few pathological states. In the absence of one of these specific conditions, present knowledge indicates that one carbohydrate is as good as another (among those commonly used for feeding); economy is therefore an important determining factor. At the present time cane sugar, the cheapest form of good carbohydrate, is the most widely used in infant feeding. Carbohydrate supplements are still generally employed in preference to unsupplemented milk formulas, because they provide a cheaper source of calories, and because unsupplemented cow's milk provides an intake of protein that is unnecessarily high, being well beyond requirements plus an adequate margin of safety.

Acknowledgment

The studies with tagged carbohydrates were carried out by a group of investigators, including Olli Somersalo, Leon Hellman, and Joseph Pescatore. Measurements of carbon-13 were made through the courtesy of Sidney Weinhouse of the Lankenau Hospital Research Institute. Measurements of carbon-14 were made at Memorial Hospital. Much valuable assistance was received from staff members of the Brookhaven National Laboratories in the preparation of the isotopic carbohydrates. The author is also indebted to I. H. Ebbs and A. L. Chute of the Hospital for Sick Children in Toronto for the privilege of studying some of their celiac patients. A complete report of these studies is in preparation. A preliminary report of part of this work was published by Somersalo (2).

A more complete discussion of the role of carbohydrates in infant feeding, with references, will be found in the excellent monograph by Lanman (1).

Literature Cited

(1) Lanman, J. T., "Modern Trends in Infant Nutrition and Feeding," New York, Sugar Research Foundation, 1952.
(2) Somersalo, Olli, *Am. J. Diseases Children*, **84**, 767 (1952).

RECEIVED November 12, 1954.

Role of Sweeteners in Food Flavor

LOREN B. SJÖSTRÖM and STANLEY E. CAIRNCROSS
Arthur D. Little, Inc., Cambridge, Mass.

The major uses of natural sweetening agents in the food and beverage industry are well understood but the subtle use of these materials in seasoning has been underemphasized. While the effect of adding small quantities of salt and monosodium glutamate to processed foods is well known, there is a lack of published information on the accomplishments of sweeteners at similar levels. Formal studies by panels of trained tasters were conducted along both fundamental and applied lines. A sweetness curve for dried corn sirup (dextrose equivalent of 42) in terms of dextrose is given and the effects of mixing solutions of dried corn sirup and sucrose are depicted. Dextrose and sucrose solutions have different taste characteristics, and sweeteners produce different effects at different concentrations. This is borne out in studies of interactions of the basic taste factors, which also explain the blending phenomenon. Sweeteners act as seasoners for fresh vegetables, canned meat products, and seafood. Seasoning action was manifested as enhancement of total flavor, blending of flavor notes, replacement or intensification of sweetness, reductions of bitterness, sourness, and saltiness, and control of flavor factors of rawness.

While both aroma and taste are obviously important in developing a food product which will supply full mouth satisfaction, inadequate attention has been given to the importance of a proper balance of sweet, salty, sour, and bitter. This subject comes within the field of seasoning. More emphasis should be placed on studies of the role of the four primary taste factors in the seasoning of foods. Commercial practice puts the principal responsibility on salt and it is well known that monosodium glutamate plays a separate and important role in seasoning.

Throughout the world, cooking practices feature the use of sweet and sour flavoring and seasoning agents. Possibly, in America, some interesting applications have been overlooked, not only of sweet and sour but also of bitter. While all of the primary taste factors commonly appear in natural foods, their actual intensities are usually readjusted in the process of cooking and seasoning. Such readjustments have a major effect on the control of unusual flavor characteristics, including aftertaste, and constitute a major step toward the creation of a blended flavor complex of high impact.

The authors have been working for many years on seasoning in connection with monosodium glutamate and have published a number of papers indicating its potentialities in seasoning, with limited coverage of the role of salt and spices. From several years of work in separate studies on the value of various sweetening agents in the flavoring of foods, the conclusion has been made that sweetness has been underemphasized in this field. There is probably a very primitive appeal connected with both saltiness and sweetness, the unconscious response being that material having these properties is probably suitable for eating. Sourness and bitterness, on the other hand, at high levels may constitute a warning against eating, but at proper levels they may contribute much to mouth satisfaction.

Fundamental Aspects

Before reporting the contributions of sweeteners to food flavor, the results of certain studies undertaken to add fundamental flavor information about the various natural sweeteners, such as their likenesses and differences, the variations in their detectability at different levels, and their action with the other basic taste factors are presented.

Cameron (2) has reported the relationships of glucose, sucrose, fructose, dulcin, and other sweet-tasting substances by plotting sweetness intensity (in terms of dextrose concentration) against concentration. Because dried corn sirup [Clintose (dextrose equivalent, 42), Clinton Industries, Inc.] has become commercially available in large quantities, its sweetness equivalency curve was determined by Cameron's method using panels of trained tasters. This is given in Figure 1, with an equivalency curve for sucrose obtained in the present work which checks that of Cameron.

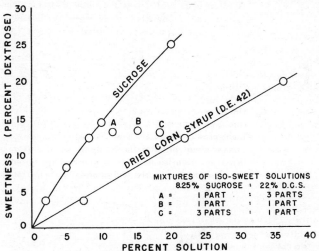

Figure 1.   Comparative Sweetness of Sucrose and Dried Corn
Sirup in Terms of Dextrose

Of even more importance is the fact that the panel found mixtures of iso-sweet solutions of sucrose (8.25%) and dried corn sirup (22%) to be slightly sweeter than either parent solution. This is represented in the figure as *A*, *B*, and *C*, referring to the following proportions by weight of sucrose and dried corn sirup solutions: 1 to 3; 1 to 1; and 3 to 1, respectively.

Additional tests showed similar trends with similar mixtures of these sweeteners at the sweetness level of 2.25% sucrose (9.25% dried corn sirup), but not at the sweetness level of 1.75% sucrose (6.75% dried corn sirup).

These findings suggest that it may be possible for a producer to enhance sweetness by adding combinations of sweeteners, provided he is working at medium levels of sweetness. In this connection, it is important to note that the just noticeable differences in sweetness vary with different degrees of sweetness. Thus, at higher concentrations of sweetness, the just noticeable difference is greater than at the lower concentrations. The tasters were easily able to distinguish between 1% increments of the dried corn sirup at the 4 to 10% levels, but not at the higher ranges. These findings check those of Lemberger (9), who studied sucrose.

Another aspect meriting consideration is the difference in the character of sweetness. While two solutions of sweeteners may be of equal intensity, it does not follow that they are indistinguishable. For example, the intensity of sweetness of a 21% dextrose solution drops off suddenly compared with a sucrose solution of equivalent sweetness (15%), which persists for a longer time, disappearing gradually. The same comparison was noted at a lower level of sweetness with solutions of 3.75% dextrose and 2.0% sucrose. However, when the sweetness of both was reduced to recognition threshold, as represented by 1.3% dextrose and 0.6% sucrose solutions, such differences were not apparent.

The preceding study again emphasizes the fact that high levels of sweeteners have different taste characteristics from low levels. In processed fruits and confections, sugar, used in quantities of 30% and above, imparts strong sweetness and this becomes a major part of flavor.

In beverages at a 10 to 12% level, sugar adds interest and boosts the flavor, while at a seasoning level, 0.2 to 1.5%, sugar may improve the flavor of food without adding conspicuous sweetness.

In focusing attention on sweetness as a single variable in seasoning, it is important to keep in mind the known ability of these taste factors to augment or suppress each other. Fabian and Blum (8) have shown that at certain levels of concentration sodium chloride tends to reduce the sourness of acids and to increase the sweetness of sugars. Acids other than hydrochloric tended to increase the saltiness of sodium chloride. Hydrochloric and acetic acids were reported to have no effect on the sweetness of sucrose, but other acids tended to increase sweetness. Sugar at some levels reduced the saltiness of sodium chloride and the sourness of acids, notably the fruit acids. In reviewing the position of salt in the food industry, Dunn (7) presents a brief review of the work of Fabian and Blum and provides some additional guidance on the respective effects of salt and sugar in the curing of meat. He also indicates that minor percentages of salt, such as 0.15, tend to increase the sweetness of beverages, whereas salt may reduce the cloying sweetness of certain confections and icings.

Studies have also been conducted in this laboratory on this subject with panels of trained tasters using the triangulation method. The findings, summarized in Table I, again emphasize the fact that the effect of one taste factor upon another is a function not only of the tastes themselves but of their concentrations.

Many flavor workers have long been interested in unraveling some of the mysteries encountered during their studies of highly blended flavors. Knowing how complicated is the chemistry of food flavor, one can conclude that any number of taste and flavor factors must be submerged within the flavor blend. Therefore, it should be possible to formulate mixtures that have no readily identifiable taste or odor characteristics. Such mixtures could be referred to as essentially "white flavor" or "white taste." As an example of this idea, so-called white taste mixture has been formulated which was designed to contain sweet, salty, sour, and bitter characteristics at threshold or near threshold values:

$0.010M$	Sucrose	1 time threshold
$0.0002M$	Citric acid	1 time threshold
$0.014M$	Sodium chloride	2 times threshold
$0.000004M$	Quinine sulfate	2 times threshold

This mixture produces an interesting sensation on the tongue, but provides almost no clues as to its actual composition. Panel members described the taste sensation as "a feeling of body but no taste."

When the concentration of individual components was increased to a level of approximately 60 times threshold weight, there was little blending, and individual taste factors could be perceived separately. The tasters, however, were still confused as to the actual identity of the components. This mixture produced most of the mouth sensations common to grapefruit juice.

Table I. Effect of Some Basic Taste Factors on Each Other

Taste Factor	Concn., %	Additive	Concn., %	Effect
Sucrose	3–10	Salt	1	Sweetness reduced
Sucrose	5–7	Salt	0.5	Sweetness augmented
Acetic acid	0.04–0.06	Sucrose	1–10	Sourness reduced
Sucrose	1–5	Acetic acid	0.04–0.06	Sweetness not affected
Sucrose	6 and above	Acetic acid	0.04–0.06	Sweetness reduced

Sweeteners as Seasonings

The next phase of the work centered on the action of sugar in various foods. Some of this has been reported elsewhere (3), but a few of the details are necessary in order to amplify the theme of this paper. The findings were obtained through the use of a panel of trained tasters employing Flavor Profile (1) techniques. Interpretations were made according to Flavor Profile concepts. These concepts maintain that the major portion of the flavor of a food is a complex blending of subthreshold quantities of flavor notes and that the identifying characteristics of the food are due to this blend and to a relatively small number of flavor notes that project beyond, at threshold intensities or stronger.

Flavor appeal rests on several psychological factors, one of which results from the judicious blending of character notes—for example, a first step in improving the palatability of foods is to cook them. In addition to altering texture and digestibility, cooking markedly changes food aroma and flavor. In general, these changes are regarded as favorable ones, since usually breadth of flavor, and thus interest, are developed and the character notes associated with rawness, which is less appealing, are suppressed and blended into the over-all flavor.

A primary example of blending was presented under the topic "white taste." In foods, however, it is not necessary or desirable to produce a perfect blend, since the principal interest factors would be lost.

A consumer usually knows what to expect in the flavor of a food and downgrades it when the flavor is unbalanced or in some manner is not representative. If, for example, the characteristics of rawness, such as earthy, starchy, bitter, and sour, are of relatively high intensity in a cooked vegetable, the vegetable will have more negative than positive flavor features. A most important means of shifting emphasis to the positive side is to add salt, along with such flavorful materials as butter and pepper and, as these present studies indicated, possibly a sweet substance.

The terms seasoning and flavoring are often used interchangeably. More exactly, they should be used according to the following definitions: A seasoning is a material which, when added to food, alters and corrects the aroma and flavor mainly by blending the character notes and augmenting total flavor impression (amplitude) without being necessarily detectable itself. A flavoring may have seasoning action, but it is recognizable in the food and adds its character to the aroma and flavor.

With this general background, some work has been carried out on the effect of low concentrations of sucrose and, in some cases, dextrose on the flavor of cooked foods which are commonly seasoned with butter, salt, and pepper. In earlier studies of various sweeteners, sweet taste was a flavoring, since it was predominant in the flavor of the strawberry preserves (10), frozen peaches (5), red raspberries, blackberries (4), and strawberries. In addition, it counteracted some of the inherent sourness and enhanced natural fruit character. The following studies were designed to test the potentialities of sweeteners as seasonings.

Fresh Vegetables. Tender, young, garden-fresh peas are succulently sweet, while too often those purchased from the local grocer lack this characteristic and taste starchy instead. When added to the cooking water of starchy flavored peas in amounts up to 4%, sucrose served to restore the desirable sweetness and at the same time controlled starchiness. This finding, of course, is not original, for many housewives put it into practice and Cruess (6) has applied it to commercially frozen peas.

Addition of the sweetener exemplified the rather simple function of replacing some of the lost natural sweetness. The best flavor balance for the peas was achieved with sugar, 1% salt, 0.01% pepper, and 2% butter.

Carrots are another vegetable which acquire less attractive flavor with increasing age. While cooking is very helpful in reducing raw flavor notes, described as earthy, metallic, terpy-green, and soapy, and also in developing sweet, buttery, and caramel notes, it does not equilibrate the flavor differences between small carrots (1⅛ inches and less in diameter) and large ones (1½ inches and greater).

The addition of 0.25% sucrose to the cooking water of large-sized carrots complemented the flavor gains attributable to cooking by enhancing carrot flavor, blending, and further reducing those flavor factors associated with rawness. Finally, the addition, just before serving, of conventional amounts of salt, pepper, and butter reduced the differences even more.

Several important experiments were devoted to corn. In all tests the kernels were removed from many cobs and mixed, in order to provide comparable samples. As much as any vegetable, corn has a complex flavor; therefore, considerable orientation work was required before testing the effect of cooking, sugar, and other additives on the flavor of fresh and aged corn. During this period, in which various increments of sugar were being checked in corn cooked in 1.5% salt solution (2 parts by weight of corn to 1 part of salt solution), it was found that additions of about 4% sugar made the samples comparable in sweetness to a popular brand of canned cream-style corn, which was frankly sweet. At 6 and 8%, sweetness was considered too high; but the addition of salt ad lib. was responsible for reducing sweetness in these samples to an acceptable level.

The taste panel described the flavor of raw, fresh corn as an unblended mixture predominantly starchy, sweet, and astringent. Cooking in tap water developed the flavor body and blended the various flavor notes. Sweetness, graininess, and butter-like notes in the cooked unseasoned corn contributed to the over-all character, while factors typical of rawness were much suppressed.

Fresh and aged corn (ears held for 3 days at room temperature) were found considerably different in flavor when cooked. Butteriness and sweetness were less in the aged corn, while earthiness, a sulfide integer, metallicness, sourness, and bitterness developed. The effect of adding 1% sucrose during the cooking of cream-style corn was to intensify natural sweetness and to reduce both sourness and bitterness. Thus, it is implied that sugar made aged corn taste more like fresh corn.

Salt, butter, and pepper not only added some of their own character, thereby increasing amplitude of flavor, but also accomplished a blending of such factors as sulfide, bitter, sour, and green vegetable. When sugar was also included, the over-all flavor was further improved, as the sweetness and the typical graininess of corn were elevated.

Tomato flavor also responded favorably to sugar. This was demonstrated in canned tomato juice, fresh tomato juice, and stewed tomatoes. The general achievements of sugar in these food were: increased blending of the flavor, reduced sourness, and decreased "green" notes. Optimum levels in the canned juice were 0.5 to 1%, while they were 2 to 3% in the freshly prepared juice. In both types sugar reduced saltiness.

In unseasoned stewed tomatoes, 2% sucrose reduced seed flavor and grassiness, augmenting sweetness and cooked flavor. An even more interesting effect was the increase of flavor amplitude. Sugar was found complementary to salt, pepper, and butter in the stewed tomatoes. Again, it was noted that sugar reduced saltiness.

Of all the vegetables reported thus far, cooked spinach possesses the fewest pleasant flavor notes. The Flavor Profile for cooked unseasoned spinach lists the following characteristics in their order of perception: bitter, sour, metallic, green (stems), earthy, and astringent. According to the authors' philosophy of flavor, 0.5% sugar performs an important service in blending these notes and in delaying the perception of bitterness and sourness. It has real, though limited, seasoning action in spinach.

One vegetable with which sugar was found incompatible is summer squash. Although it did increase the natural sweetness of the boiled squash, 0.25% sucrose upset the delicate flavor balance of the squash. Especially noticeable was the increase in metallic and vegetable-sour notes.

Canned Meat Products. It has been long recognized that sugar plays an important part in the development of flavor in cured meats. The earlier observations on the reduction of saltiness by sugar led to a checking of various additions of sugar in a canned corned beef hash, a moderately salty food. The panel found that all samples prepared with added sugar, ranging from 0.25 through 4%, tasted less salty. Levels of 1% and above were rejected because of oversweetness. With 0.25 and 0.5% added sugar, the samples were considered highly acceptable because of increased total flavor and more tolerable saltiness.

The flavor of freshly prepared and canned beef stew using 0.25 and 0.5% of sugar was found to be improved. In the blended flavor, saltiness, a starchy note, and onion flavor were reduced.

Seafoods. Very striking results were obtained in some seafood preparations. A canned Manhattan-style clam chowder, heated with 0.25% sucrose, was reported as having increased clam flavor, while the other notes, especially starchy and peppery, were reduced. Similarly, the addition of 0.5 to 0.6% dextrose enhanced the clam flavor of a brand of New England clam chowder. The cooked milk flavor and starchiness of the potatoes were blended and saltiness was reduced.

Sucrose and dextrose at the same levels as for the chowders were added to canned lobster. Both sweeteners reduced the briny-saltiness and sour notes. Dextrose surpassed sucrose by making the meat taste more like fresh lobster.

Two aromatic notes are outstanding in fresh shrimp. One is an amine type and the other resembles the starchlike note observed in hominy and tortillas. The latter is referred to as an indolic note. Chilled fresh shrimp which had been boiled in salted waters containing 2 and 3% sucrose were found to be almost lacking in these amine- and indole-type notes, which some people regard as detracting features. Saltiness was also reduced. Blending of the flavor notes and increase in flavor body were apparent. Similar observations were also made when 5 and 10% sucrose were added to the cooking waters. In addition, the shrimp meat tended to resemble lobster or crabmeat.

It has been shown that low levels of sweeteners (dextrose and especially sucrose) can be used as seasonings. In nearly every food tested, sweeteners used at low concentration improved the blending of the various flavor notes and in many instances this was accompanied by an increase in total flavor. Almost invariably, when natural sweetness has been reduced, sucrose enhanced it. Usually, undesirable levels of sourness and bitterness were diminished, and very frequently saltiness was reduced.

Conclusions

In view of the great complexity of variables existing in most flavor systems, the readily controlled primary taste factors would seem to warrant a great deal of separate attention. The effect of adding small amounts of salt and monosodium glutamate to processed foods is quite well known, but there is a definite lack of published information on what can be accomplished by the addition of sweetening agents at similar levels. There is an even greater lack of information on sour and bitter.

Confusion of the taste perception mechanism by the competition of sweet, salty, sour, and bitter factors may seem to be hopelessly complicated. In the authors' experience this is merely a part of the normal phenomenon of flavor blending. A better understanding will be reached when more studies such as those of Fabian and Blum (8) are carried out in actual food and beverage media.

The blended complex of primary taste factors might be called the platform on which the Flavor Profile is built and, as such, it plays a fundamental role in developing interesting and mouth-satisfying flavors.

Literature Cited

(1) Cairncross, S. E., and Sjöström, L. B., *Food Technol.*, **4**, 308–11 (1950).
(2) Cameron, A. T., Sugar Research Foundation, *Sci. Rept. Ser.*, **9**, 62 (1947).
(3) Caul, J. F., *Sugar Molecule*, **5**, No. 2 (1951).
(4) Caul, J. F., and Sjöström, L. B., *Quick Frozen Foods*, **13**, 59–61 (1952).
(5) Caul, J. F., Sjöström, L. B., and Turner, W. P., *Ibid.*, **12**, 54–8 (1950).
(6) Cruess, W. V., *Fruit Products J.*, **29**, Sect. 1 (1949).
(7) Dunn, J. A., *Food Technol.*, **1**, 415–20 (1947).
(8) Fabian, F. W., and Blum, H. B., *Food Research*, **8**, 179–93 (1943).
(9) Lemberger, F., *Arch. Pflügers ges. Physiol.*, **123**, 293–311 (1908).
(10) Swaine, R. L., *Food Technol.*, **4**, 291–2 (1951).

RECEIVED April 7, 1953.

Sugars in Human Nutrition

ROBERT C. HOCKETT

72 Howell Ave., Larchmont, N. Y.

Most of the world's population has been underfed during the entire course of history. Even now an adequate diet for all would require a tremendous increase in food supply. Science probably has means to meet the need, if political and economic conditions permit their application in time. All indications are that carbohydrates in general and sugars in particular will be depended upon even more heavily in the future because of their high agricultural yields. Hence the physiological effects of sugars are of prime importance. The fate and function of sugars in the body are traced with consideration of the mechanism of absorption, storage, caloric requirements, homeostatic regulation in blood, influence on fat and protein metabolism, and utilization by muscle and brain, and in detoxification mechanisms. Certain physiological differences between sugars such as glucose and fructose are considered with particular reference to use in handling diabetics, psychotics, and the nutrition of the aged.

The preceding papers have discussed the effects of carbohydrates, especially sugars, in a number of types of food products. The functions considered were largely those of a chemical, physical, or mechanical nature such as will affect the color, texture, flavor, and other qualities of the numerous kinds of foods of which sugars are an essential ingredient.

It hardly seems proper to close this symposium, however, without some consideration of the place held by sugars among the general food resources of the human race, and the role of the sugars in human nutrition.

Growth of World Population

The outstanding sociological fact of our times is the tremendous rate of growth of world population (7, 14, 19, 20, 67). In 200 years it has tripled, growing from 875,000,000 to about 2,400,000,000. At present it is increasing by 26,000,000 or 1.2% per year.

Population in India grew from 250,000,000 in 1870 to 400,000,000 in 1945; in Java population tripled between 1860 and 1930 (20). Japan grew from 30,000,000 in 1850 to 45,000,000 in 1900 and 83,000,000 in 1950 (7, 67). Even in the United States, population growth has astonished the experts. Some 19,000,000 were added between 1940 and 1950. At 156,000,000 in 1952, we were at least 10,000,000 ahead of the level that had been predicted for that year during the thirties (19).

It has been estimated that of all human beings who ever lived, one out of twenty is living today!

Moreover, of all the people who ever lived on this earth, it is probable that the great majority were hungry and malnourished during most of their lives! Even just before World War II, two thirds of the world's people were undernourished all the time. To give every man, woman, and child in the world the barely adequate allowance of 2600 calories a day would require an immediate doubling of the total world food supply (24).

Yet, since the war, and despite the attack upon food problems on a global scale by the Food and Agricultural Organization of the United Nations, there has been a net loss of per capita food supply. While production has gone up 9%, population has increased 12% *(22)*. The hungry are hungrier and the desperate more desperate than ever. What counts, of course, is not the absolute quantity of food but the ratio of food to population. While this ratio declines or remains too low, there are certain to be unrest, revolution, and war.

Science can accomplish many things by way of increasing the food supply, but time, capital, and a suitable political climate are required. The burning question is whether circumstances will permit improvement fast enough for food supplies to catch up with population and gain to the extent of raising the critical per capita ratio significantly. Otherwise the political situations will become still worse before they improve.

Some of the measures by which science and engineering can be expected to attain this goal, if given the opportunity, are listed in Table I *(41, 55, 56)*.

Table I. Ways of Increasing Food Supply

A. Bringing new lands under cultivation
   1. Arctic agriculture
   2. Cultivation of muck soils
   3. Irrigation of arid lands
   4. Extension of tropical agriculture
   5. Draining of swamps
   6. Reclaiming of nonproductive soils by adding deficient elements (Australia)
   7. Making lands habitable by elimination of dangerous pests (tsetse fly)
B. Increasing production of crops per acre
   1. Breeding of new plant varieties
   2. Use of more and better fertilizer
   3. Improvement of tools and machinery
   4. Better chemical control of insects and fungi
   5. Better chemical control of weeds
   6. Introduction of new plants
C. Increasing production of animal foods
   1. Breeding of more efficient strains
   2. Prevention of losses by disease
   3. More scientific feeding to obtain the same meat and milk with less feed
   4. Use of hormones, antibiotics, and synthetic feed ingredients to speed growth
   5. More use of agricultural wastes for feeding stock
D. Increased use of fish and other sea products
E. Synthesis of vitamins and amino acids, to fortify foods that are abundant but inadequate nutritionally
F. Production of stock and human food by microbiology (yeasts, bacteria, and algae)
G. Direct or indirect conversion of nonnutritive agricultural residues into food—e.g., hydrolysis of cellulose

If an economic method of salvaging sea water for use in irrigation could be perfected, the potential sociological effects might surpass those of the discovery of America or the Industrial Revolution!

The alternative to increasing food supply, of course, is control of the rate of population growth *(47)*. Many serious scientists believe that birth control will be required along with all possible efforts to produce more food, if a balance is to be obtained soon enough to save the situation. The governments of India, Japan, and Jamaica have inaugurated programs of population control. The Netherlands and Scandinavia have operated birth-control clinics for many years.

In the United States, we feed 60% of all our grain to livestock, and derive 25 to 35% of our dietary calories from foods of animal origin. This assures us, as a people, of a relatively liberal supply of high quality protein, which is generally the scarcest and most expensive part of the diet. In Asia, virtually no grain is fed to animals. Because of population pressures, the people have to eat all the grain themselves and use livestock, principally swine and poultry, as scavengers to produce human foods from garbage and offal. The Asians derive only about 3% of their calories from foods of animal origin. Their protein intake is probably close to the margin of adequacy, with beans and rice supplying the main bulk *(11)*.

No doubt we Americans could adjust ourselves to a growing population for a long time by increasing our direct consumption of grains at the expense of meat. The time is probably at hand when life and health can be maintained satisfactorily on vegetable foods alone. Such plants as soybeans, chick peas, and grasses could be grown for their relatively adequate protein and used with supplements of synthetic vitamins, particularly B$_{12}$, and perhaps with synthetic supplements of the scarcer essential amino acids. Though such changes are perfectly feasible, none of us contemplate them with any pleasure. We all hope that science will forestall any such necessity for a long time (11).

Relation of Sugar to Total Food Supply

Our main concern at the moment is to show the relation of carbohydrates in general and of sugar in particular to this problem of the total food supply available to mankind. It is probably fair to say that a very large proportion of all the food energy used by all the humans who ever lived was derived from carbohydrate. It may well be that our primitive ancestors once lived largely on wild game and hence consumed a diet chiefly of meat. There have also been pastoral or nomadic societies living mainly upon the products derived from domesticated animals which had to be followed in their migrations from one pasture to another. A few modern counterparts of these primitive societies survive; among these are certain Eskimos, the Turkana of Africa, and a few Siberian nomads (34). Nevertheless, when the race embarked upon tilling of the soil, carbohydrates became a predominant component of the diet. Indeed, it was carbohydrate foods, through agriculture, which permitted civilization as we know it to come into existence and which made city life possible. Among crowded contemporary oriental populations, carbohydrate now composes 85% of the food.

Table II gives the relative proportions of the total diet taken as fat and as carbohydrate in several parts of the world as shown by Himsworth (30) in 1936.

Table II.   Fat and Carbohydrate in Diet

	Percentage of Calories	
	From fats	From carbohydrates
United States	36.1	50.8
Holland	35.5	53.0
England and Wales	32.0	58.0
Scotland	28.3	57.2
Italy	18.2	65.3
Japan	4.7	85.0

These figures are probably accurate enough to sustain the thesis that the fat-carbohydrate ratio is an index to the standard of living, dietetically speaking.

It is possible that we are not yet perfectly adapted to such a high carbohydrate diet. Some students of tooth decay point out that dental caries is rare among primitive Eskimos and the meat eaters of the Great Rift Valley in Africa. Many seem to feel that the human race made a mistake in giving up its primitive, carnivorous diets and that dental ailments are the price we have paid for adoption of carbohydrate foods (49). However this may be, all indications are that carbohydrates are here to stay. Indeed, unless science can provide some alternative that is not yet in sight, the growth of even United States population may be expected eventually to force us by stages up toward the 85% carbohydrate level of the orientals.

There is a considerable difference in the productivity of the common food crops (Figure 1). Soybeans produce about 1,500,000 calories per acre. Rice does slightly better and Irish potatoes approach 2,000,000. Sweet potatoes produce 2,750,000 and corn averages more than 3,000,000. Both sugar beets and sugar cane, however, far surpass all other crops commonly grown on any scale for food. The average sugar acre yields more than 7,500,000 calories a year without consideration of by-products such as molasses, bagasse, beet pulp, or tops. This is more than double the score of any other major agricultural crop (31).

Of course, food values cannot be measured in calories alone. Nevertheless, both beets and cane must be classed as extremely efficient converters of solar energy into food. This fact alone is sufficient to justify classifying sugar as a major human resource. It even suggests that extension of sugar acreages will probably

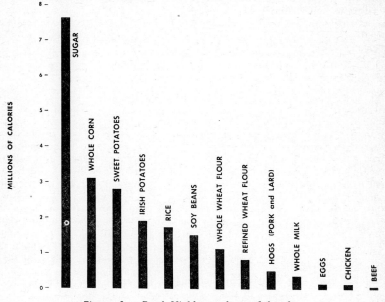

Figure 1. Food Yield per Acre of Land

provide one means of increasing global food supplies and conserving space needed for necessary but less efficient production of high quality protein. If global planning develops beyond its present embryonic state, as we must hope it will, sugar is capable of playing a very significant part, assuming that it is used intelligently. Natural forces which would be difficult to control in any case seem destined to increase the over-all per capita consumption of this product. An increase in sugar consumption has accompanied rising living standards in practically every country.

Physiological Behavior and Function of Sugars

Starches and dextrins are very widely disseminated among grains, vegetables, and legumes. They compose about one half of our American carbohydrate intake (68). Sucrose, or common sugar, is a constituent of all plants containing chlorophyll (12), although it is separated commercially chiefly from beets and cane. As a constituent of fruits and vegetables as well as in the form of refined sugar, molasses, and sirups, sucrose probably constitutes about one quarter of the carbohydrate we consume. Glucose (dextrose) and fructose (levulose) are formed from sucrose by hydrolysis, and are generally present along with sucrose in fruits and vegetables. Dextrose from starch hydrolysis is also an article of commerce in the crystalline form and in sirups. These two simple sugars together provide about 10% of the carbohydrate consumed. Lactose or milk sugar provides another 10%. Glycogen of meats and fish constitutes an almost negligible percentage, leaving about 5% to be accounted for by indigestible forms of carbohydrate such as cellulose, pectins, pentosans, hemicelluloses, and the like (68).

In a broad way, all the digestible carbohydrates perform a similar function in the body, but differences among them are significant medically and are now receiving considerable study.

Starch and sugar (sucrose) are both essentially plant products. They are not ordinarily found in the animal body proper—that is, outside the alimentary canal—nor are they produced by animals. Nevertheless, the human body is remarkably well adapted to handling both of these materials as foods, breaking them down and converting them into the substances needed in its own physiology (70). This adaptation would seem to reflect the millennia during which starch and sugar have been major constituents of our diets.

Assimilation of Carbohydrates

In the mouth, a salivary amylase is present which begins digestion of starch, glycogen, and dextrins into maltose. Sucrose is unaffected by saliva but may be attacked by bacteria present in the mouth (62, page 6). Starch digestion is rarely completed there, however, as swallowing is very likely to remove the food before the process is finished. In the stomach, hydrochloric acid stops action of salivary amylase. This acid is strong enough to attack sucrose rather rapidly, inverting it to glucose and fructose. Again, however, the stomach is likely to empty its contents into the small intestine before the process is complete. The latter is distinctly alkaline and in turn stops the acid action; intestinal enzymes are present, capable of continuing the breakdown of starch, maltose, sucrose, and lactose. Most of the digestible carbohydrates are thus reduced in the three organs to the simple sugar or monosaccharide stage. In this form they are absorbed into the portal vein and transported to the liver (62, page 6).

Undigestible carbohydrates, or those which escape digestion, pass into the large bowel, where they are exposed to the profuse and various bacterial types that normally inhabit that region. A certain amount of additional monosaccharide may be formed there and perhaps absorbed.

All monosaccharides are believed to be capable of absorption by a simple diffusion which will increase in speed with increasing concentration of the material in contact with the membrane. Three of the most important food monosaccharides are also absorbed by some other mechanism which is specific in character and is scarcely affected by the concentration of the sugar in question. Galactose is most rapidly absorbed of these three; glucose is a close second, and fructose third, in the ratio 110:100:43. Mannose and the pentoses, which are absorbed only by diffusion, have an absorption rate of 9 on the same scale. While these quantitative measurements were made on rats, a similar situation holds for man at least qualitatively (17, 27).

As the inorganic phosphate content of the blood drops during absorption of these sugars, there is a theory that the specific-rate absorption of glucose, galactose, and fructose involves a temporary combination with phosphate. The hypothesis is supported, though not proved, by the fact that several enzyme poisons can lower the rate of absorption of these sugars (69).

Absorption naturally is affected also by the condition of the intestinal wall, by the behavior of the endocrine glands and by the adequacy of vitamin supply, especially of the B complex (62, page 8).

In the liver, all monosaccharides that can be utilized at all are converted into glycogen, a polysaccharide related chemically to starch. The liver is one of the main carbohydrate storage depots of the body, though its capacity is limited. A full-grown man can store about 108 grams of glycogen in the liver. The muscles, taken together, can store about 245 grams or half a pound. Stored glycogen is reconverted to glucose and fed as needed into the blood stream. Rather sensitive regulatory machinery maintains the blood sugar level within rather narrow limits. The actual amount of glucose in the total blood at any time is only about 17 grams or little more than 0.5 ounce.

The body's total carbohydrate store is thus only about 370 grams or less than 14 ounces (62, page 10).

	Grams
Muscle	245
Liver	108
Blood	17
	370

This is really an extraordinarily small amount of carbohydrate storage. The blood sugar level must always be maintained above 40 to 50 mg. per 100 ml. to keep the brain and central nervous system functioning *(33, 50).* If readily available sugar falls below the critical level for even a few minutes, convulsions and death will result. Fat, unlike carbohydrate, can be stored in almost unlimited quantity, as many of us have observed from our own experience; protein makes up about 75% of the body's dry weight. This protein really is the body proper and can function in many different ways, but it can also serve as a food storage depot.

An average man, even when engaged in sedentary occupations, will burn up an ounce of sugar per hour or the equivalent. The maximum bodily reserve of carbohydrate would therefore supply the caloric needs for less than 14 hours.

Regulation of Fat Metabolism

These facts suggest that the normal situation is a frequent replenishment of carbohydrate supply by new food intake. Yet consumption of more carbohydrate at any one time than the liver, muscles, and body fluids can store with their total capacity for 14 ounces of glycogen will not further augment the carbohydrate reserve. Instead, the surplus will be converted and stored as fat.

With the nervous system so critically dependent upon a constant sugar supply and with the small reserve of carbohydrate ordinarily present in the body, life would hang by a slender thread indeed, unless blood sugar could be derived from sources other than glycogen when necessary. Fortunately, the breakdown of fat can refurnish some carbohydrate through the glycerol part of the molecule. In general, however, the fats, even though originally formed from carbohydrate, do not produce very much carbohydrate when they are broken down again. The fatty acids mainly follow another route of degradation to form chiefly β-hydroxybutyric and acetoacetic acids. This transformation occurs primarily in the liver, but the products, often called "ketone bodies," are transported throughout the body, where the muscles complete the oxidation with release of energy *(62,* page 112). The peripheral tissues can dispose of ketone bodies efficiently as a rule, so that little β-hydroxybutyric acid or acetoacetic acid is ordinarily detectable in the blood.

However, glycogen in the liver somehow acts as a regulator of fat degradation. Once the glycogen reserve has been exhausted, the production of ketone bodies goes haywire, production is enormously increased so that the peripheral tissues become saturated, and these compounds accumulate in the blood *(62,* page 121). Severe "ketosis" may cause nausea, vomiting, and other distressing and even dangerous symptoms *(43).*

Thus, one function of carbohydrate is the regulation of fat metabolism. The total picture of carbohydrate-fat interrelationships is a most remarkable one, and an understanding of it is extremely important in the management of reducing diets and diabetics. We can summarize the situation by saying that a constant surplus intake of carbohydrate over the body's needs will never produce more than a small store of carbohydrate, but will then add to the store of fat and in time may build up a large reservoir of the latter. A reversal of the process, by total abstinence from food, will quickly use up the small carbohydrate store and then degrade fat in a disadvantageous and unregulated manner. The desirable situation in a reducing diet is to dribble in enough carbohydrate at frequent intervals to keep the liver supplied with glycogen for regulating this fat metabolism. At the same time, this carbohydrate feeding should never at any time be liberal enough to saturate the glycogen stores and permit any part of the intake to spill over into the fat reserve. It is obvious, of course, that direct intake of fat must be drastically curtailed during a regimen designed to reduce the body's store of fat.

Reducing Diets

Thus, it is correct to say that carbohydrate is very important in reducing diets, though the amounts have to be regulated. There is also another reason why carbohydrate is important in such diets. Evidence has accumulated recently that both certain animals and humans with a pronounced tendency to become obese are fat eaters *(39, 40).* In other words, they tend to select diets high in fat rather than high in carbohydrate. For some reason, the consumption of fat by these animals and humans does not seem to turn off the machinery that is responsible for the appetite. On the other hand, sugar seems to "pull the trigger" very promptly and thus to shut off the psychological or physiological impulses that make the creature "feel hunger" or, at any rate, consume food. In many cases a little sugar, given 15 or 20 minutes before a meal, will cause the animal (or person) to eat less than he would have eaten without this preparation, thus producing a net reduction in total caloric intake. This is perhaps nothing more, really, than the old observation that sugar "kills the appetite" or has "high satiety" value, backed up with new and more exact observation. This effect apparently can be used in a practical way in reducing diets.

The protein of body substance can serve as a food reserve if necessary. In general, proteins are relatively scarce and expensive. Their characteristic ingredients are, of course, the amino acids, eight of which are classed as strictly essential for humans. Fifteen others are at least useful under proper conditions (54). The amino group nitrogen is common to all.

When glycogen stores become depleted, tissue protein can be mobilized to furnish sugar. Degradation of proteins to amino acids will occur first, followed by a process of deamination to keto acids. Some of the latter can be built up into glucose to maintain blood-sugar level, while the nitrogen is excreted, mainly as urea. This nitrogen excretion provides a clue as to what is happening to proteins in the body (52).

In total starvation, glycogen is first depleted, then much of the stored fat along with some protein, and finally tissue protein, until the exhaustion of some essential organ causes death. Just before death a flood of nitrogen excretion usually occurs, perhaps reflecting a last desperate and inefficient destruction of tissue to maintain the critical blood sugar.

When protein is taken by mouth, it is always degraded to its constituent amino acids in the small intestine, and these smaller fragments are carried to the liver. There, these amino acids approach a fork in the road. On the one hand they can be recombined to form the characteristic protein patterns of the person who has digested the food in question. On the other hand, they can lose their nitrogen and go to the formation of glucose. This glucose may, like any other, be deposited as glycogen. If glycogen depots are filled, it can be transformed into fat (45). These changes are essentially irreversible. Fat, glycogen, and glucose are not significantly transformed back into amino acids or protein, once the deamination or nitrogen loss has occurred. To be sure, a transfer of amino nitrogen may occur from one amino acid to a ketone with formation of a new amino acid. In such cases there is simultaneous destruction of the old amino acid, so that only transformation occurs, with no net gain (15).

Carbohydrates as Body Fuel

When glycogen stores are low, the condition is favorable for transformation of protein into glucose. If carbohydrates are eaten along with protein, however, they are used for glycogen formation in preference to the amino acids (65). Thus, carbohydrates spare proteins for their unique functions in the building and repair of bodily tissues. This is a matter of some consequence because of the relative scarcity and high cost of proteins as ingredients of the diet. Physiologically, it is possible to live almost exclusively on proteins, because these can not only build and repair tissue but also produce glycogen and fat (35, 36, 63, 66). In this sense, proteins are the most versatile of the foods. Economically, however, the use of protein as fuel is a good deal like burning fine furniture in the fireplace. This sort of thing has been done in human experience, under special circumstances, but is not contemplated in planning. The normal protein proportion of the diet is only 10 to 12% expressed as percentage of the total caloric content. This protein must also be qualitatively adequate in its content of essential amino acids. Even the so-called "high-protein" diets are only relatively high and as a rule contain substantial proportions of carbohydrate (1, 48).

Such high-protein diets are often prescribed for weight-reduction purposes. They have certain advantages. High protein provides assurance that wear and tear of bodily tisues will be amply provided for, while starvation ketosis during rapid fat breakdown will be prevented by the glycogen formation mechanism described. High protein also stimulates the metabolic rate, keeping the bodily fires burning brightly, which is an advantage. Where meat protein is the backbone of the diet, bulk is relatively great, and most people find it pleasant to take and relatively easy to keep track of. The main disadvantages are high cost and physiological wastefulness (63, 74).

There are cases on record where the protein allowance in certain institutional homes and schools was adequate in itself, or should have been, but failed to meet the nutritional needs of children because they did not get enough carbohydrate with it to prevent a significant part of it from being burned up as fuel. It is paradoxical but true that an addition of sugar alone to such a diet may increase its effective protein value through the "sparing" effect (37).

As fuel for muscular exercise, the carbohydrates have been shown to be the most direct and efficient materials. The effects are especially marked during prolonged or severe exercise. Fats and proteins are used very well, but less directly and with some loss of energy in the transformations involved (25).

The brain and nervous system appear to use only sugars directly as fuel, which explains why a minimum level of glucose in the blood is essential to life (33, 50). The heart muscle, on the other hand, can use glucose, acetic acid, lactic acid, and products formed from fat, though lactic acid (of carbohydrate origin) is its preferred fuel. If muscular damage has occurred in the heart, this organ performs better when supplied liberally with fuel. Cardiac patients are often given frequent small feedings to keep blood sugar at a relatively high level. Any carbohydrate that furnishes glucose to the blood can be used, and this blood sugar presumably keeps up the lactic acid supply (8, 73).

The glycogen store normally present in liver also plays a role in elimination of poisons such as carbon tetrachloride, chloroform, p-aminobenzoic acid, sulfanilamide, alcohol, arsenic, and bacterial poisons (10, 26, 28, 32, 51, 60, 61, 75). It probably plays a role also in regulating steroid hormone metabolism and in eliminating excess sex hormones. The elimination of substances that otherwise would be carcinogenic may turn out to involve the liver and liver glycogen (62, page 14).

The common anesthetics are poisonous to a degree. A good glycogen store ensures against liver damage and also helps in recuperation from surgical shock. It has therefore become virtually routine to administer sugars by the intravenous route prior to operations.

All sugars and other carbohydrates that are utilized by the body form glycogen. Despite certain claims, there is no clear evidence that more than one type of glycogen exists. Hence, all carbohydrates might be expected a priori to behave alike once they have reached the common pool represented by glycogen. In a sense, these considerations may have delayed recognition of some significant differences that exist among sugars and promise to be useful in medical practice.

Fructose

As a part of the sucrose molecule and a number of polysaccharides, fructose, commonly called levulose, has been consumed by people for thousands of years. Only recently, however, has pure fructose been available in a form suitable for careful biochemical and physiological studies and in sufficient quantity to encourage sustained attention.

Fructose crystallizes well when pure. It is much more soluble and considerably sweeter than common sugar. It is a "physiological" sugar in every sense of the word. Now that we have analytical methods capable of distinguishing fructose from glucose certainly, it has been shown that fructose is normally present in human blood (71). In embryonic and newborn infants, the level is much higher than in adults (5). Fructose is also the predominant sugar of seminal fluid and its concentration has an important effect upon the motility of spermatozoa (38).

Fructose is less rapidly absorbed through the small intestine than glucose or galactose. Despite this handicap, fructose taken by mouth registers an effect upon the carbon dioxide content of expired air considerably faster than glucose (21, 53). It also produces much more lactic acid, pyruvate, citrate, and glutarate than glucose (64).

The handicap of slower intestinal absorption is overcome when fructose is given by vein. Uptake of fructose by the liver is very rapid *(72)* and this sugar has been shown to produce glycogen much faster than glucose *(18)*.

As might be expected from these facts, fructose can be administered rapidly without producing a very great rise in the blood sugar levels. It is even reported that fructose administration speeds up the disposition of glucose, causing a drop in high blood-glucose levels when administered *(23)*.

A more marked protein-sparing action has been reported for fructose as compared to lactose, glucose, sucrose, or dextrins *(2, 44)*. While both glucose and fructose speed alcohol metabolism, fructose has the greater effect *(13)*.

Fructose has been reported by many observers to be handled more nearly normally by diabetics than starch or glucose. Fructose gives a greater rise in respiratory quotient and produces the normal pyruvate rise, providing that ketosis has not developed. Even with ketosis present, fructose is absorbed by the livers of diabetics at a normal rate *(42)*.

Diabetic rats form fats in normal fashion from acetate and lactate when fed fructose, whereas glucose-fed diabetic rats experience a marked depression of fat formation *(6)*.

These differences in carbohydrate handling are both genuine and distinct. Nevertheless, it is not yet altogether clear how they are to be exploited practically in management of diabetics. Older experiments upon feeding of fructose to diabetics were reported not to produce sustained improvement *(46)*, but newer biochemical knowledge suggests that clinical studies now need to be repeated with more complete controls.

Many victims of mental disease are said to show disturbances of carbohydrate metabolism. Indeed, the physiological and biochemical study of mental diseases is in its early infancy. Preliminary findings suggest that the time is ripe for full-scale attack upon mental diseases by the techniques of these sciences. The insulin shock treatment represents only a first crude application of physiological techniques to therapy of such conditions. More effective methods may be expected to evolve out of this beginning.

In general, psychotics are reported to handle fructose in a more nearly normal fashion than glucose. What this basic fact may mean eventually in practical treatment of such cases cannot yet be estimated *(4, 29)*.

Since fructose is reported to speed up disposition of glucose, it has been suggested that fructose might heighten insulin action in its initial stage, conserving the expensive hormone, and perhaps assisting in treatment of insulin-resistant diabetics or psychotics *(4, 29)*.

The increasing length of our average life span is attracting more study to geriatrics. The altered nutritional requirements of older people and their changes in physiological behavior with aging are subjects of tremendous and growing social importance. It has been shown that glucose tolerance generally declines with age *(57)*. Again the utilization of fructose appears to be more nearly normal or less affected by senescence *(3)*. Whether this fact will prove important in forestalling some of the effects of old age, we do not yet know.

Tooth Decay

In a broad way tooth decay appears to be associated with carbohydrate diets and to be insignificant among those (now exceptional) humans who consume little but meat. Soluble sugars are not by any means wholly to blame for the local mouth conditions unfavorable to teeth. There is ample evidence that many forms of starch have similar effects *(9, 16)*.

There are enough examples of animals and humans who can handle high carbohydrate diets without tooth damage to prove that a mechanism of resistance exists *(58)*. Experimental evidence is accumulating to show that resistance can be altered and controlled, particularly by attention to pre-eruptive feeding *(59)*. The fluoride story is the best known part of this development, but only one part. Such findings, early though they are, provide substantial hope of a human race eventually free from decay of the teeth. This is fortunate, for the carbohydrates form the backbone of our human diet throughout the world. All present indications are that we must depend even more heavily upon sugars and starches in the future.

Literature Cited

(1) Albanese, A. A., "Protein and Amino Acid Requirements of Man," p. 116, New York, Academic Press, 1950.
(2) Albanese, A. A., Felch, W. C., Higgons, R. A., Vestal, B. V., and Stephanson, L., *Metabolism Clin. and Exptl.*, **1**, 20 (1952).
(3) Albanese, A. A., Higgons, R. A., Orto, L., Belmont, A., and Dilallo, R., *Ibid.*, **3**, 154 (1954).
(4) Altschule, M. D., private communication.
(5) Bacon, J. S. D., and Bell, D. J., *Biochem. J.*, **42**, 397 (1948).
(6) Baker, N., Chaikoff, I. L., and Schusdek, A., *J. Biol. Chem.*, **194**, 435 (1952).
(7) Balfour, M. C., Evans, R. F., Notestein, F. W., and Taeuber, I. B., "Public Health and Demography in the Far East," New York, Rockefeller Foundation, 1950.
(8) Best, C. H., and Taylor, N. B., "Physiological Basis of Medical Practice," p. 267, Baltimore, Williams & Wilkins, 1943.
(9) Bibby, B. G., Goldberg, H. J. V., and Chen, E., *J. Am. Dental Assoc.*, **42**, 491 (1951).
(10) Bollman, J. L., *Arch. Pathol.*, **29**, 732 (1940).
(11) Brody, S., *Federation Proc.*, **11**, 667, 689 (1952).
(12) Calvin, M., *J. Chem. Educ.*, **26**, 639 (1949).
(13) Carpenter, T. M., and Lee, R. C., *J. Pharmacol. Exptl. Therap.*, **60**, 286 (1937).
(14) Carr-Saunders, A. M., "World Population, Past Growth and Present Trends," Oxford, Clarendon Press, 1937.
(15) Cohen, P. P., "Transamination," in "Symposium on Respiratory Enzymes," p. 210, Madison, University of Wisconsin Press, 1942.
(16) Constant, M. A., Phillips, P. H., and Elvehjem, C. A., *J. Nutrition*, **46**, 271 (1952).
(17) Cori, C. F., *J. Biol. Chem.*, **66**, 691 (1925).
(18) *Ibid.*, **70**, 577 (1926).
(19) Davis, J. S., "The Population Upsurge in the United States," San Francisco, Stanford University Press, 1949.
(20) Davis, K., "The Population of India and Pakistan," Princeton, Princeton University Press, 1951.
(21) Deuel, H. J., Jr., *Physiol. Rev.*, **16**, 173 (1936).
(22) Dodd, N. E., "The Work of the FAO," Report of the Director General, Rome, Food and Agricultural Organization of the United Nations, 1951.
(23) Fletcher, J. P., and Waters, E. T., *Biochem. J.*, **32**, 212 (1938).
(24) Food and Agricultural Organization of the United Nations, "World Food Appraisal for 1946–47," 1946.
(25) Gemmill, C. L., *Physiol. Rev.*, **22**, 32 (1942).
(26) Glahn, W. C., Flinn, F. B., and Keim, W. F., *Arch. Pathol.*, **25**, 488 (1938).
(27) Groen, J., *J. Clin. Invest.*, **16**, 245 (1937).
(28) Heller, C. G., *Endocrinology*, **26**, 619 (1940).
(29) Henneman, D. H., private communication.
(30) Himsworth, H. P., *Clinical Science*, **2**, 117 (1935).
(31) Hockett, R. C., *J. Calif. State Dental Assoc.*, **26** (Supplementary Issue), 76 (1950).
(32) Holmes, E. G., *Physiol. Rev.*, **19**, 439 (1939).
(33) Jensen, H. F., "Insulin: Its Chemistry and Physiology," London, Oxford University Press, 1938.
(34) Lewis, K., *Trans. N. Y. Acad. Sci.*, **10**, 245 (1948).
(35) Lieb, C. W., *J. Am. Med. Assoc.*, **93**, 20 (1929).
(36) McClellan, W. S., *J. Am. Dietet. Assoc.*, **42**, 397 (1948).
(37) Mack, P. B., "The Value of Sugar for Children on a Submarginal Caloric Intake," New York, Sugar Research Foundation, Inc., 1950.
(38) Mann, T., *Biochem. J.*, **40**, 481 (1946).
(39) Mayer, J., *Physiol. Rev.*, **33**, 472 (1953).
(40) Mayer, J., Dickie, M. M., Bates, M. W., and Vitale, J. J., *Science*, **113**, 745 (1951).
(41) Maynard, L. A., *Federation Proc.*, **11**, 675 (1952).
(42) Miller, M., Drucker, W. R., Owens, J. E., Craig, J. W., and Woodward, H., *J. Clin. Invest.*, **31**, 115 (1952).
(43) Mirsky, I., and Nelson, N., *Am. J. Diseases Children*, **67**, 100 (1944).
(44) Munro, H. N., *Physiol. Rev.*, **31**, 449 (1951).

(45) Newburgh, L. H., in "Clinical Nutrition," ed. by Jolliffe, Tisdale and Cannon, pp. 731 ff., New York, Hoeber, 1950.
(46) Noorden, C. von, "Die Zuckerkrankheit und Ihre Behandlung," 5th ed., Berlin, A. Hirschwald, 1910.
(47) N. Y. Acad. Sci., *Ann. N. Y. Acad. Sci.*, **54**, 729 (1952).
(48) Patek, A. J., Jr., and Post, J., *J. Clin. Invest.*, **20**, 481 (1941).
(49) Price, W., "Nutrition and Physical Degeneration," New York, Hoeber, 1939.
(50) Quastel, J. H., *Physiol. Rev.*, **19**, 422 (1939).
(51) Quick, A., *Am. Rev. Biochem.*, **6**, 291 (1937).
(52) Rapport, D., *Physiol. Rev.*, **10**, 349 (1930).
(53) Root, H. F., Stotz, E., and Carpenter, T. M., *Am. J. Med. Sci.*, **211**, 189 (1946).
(54) Rose, W. C., Haines, W. J., Johnson, J. E., and Warner, D. T., *J. Biol. Chem.*, **148**, 457 (1943).
(55) Russell, J., *Nature*, **164**, 379 (1949).
(56) Salter, R. M., *Science*, **105**, 533 (1947).
(57) Schneeberg, N. G., and Finestone, I., *J. Gerontol.*, **7**, 54 (1952).
(58) Shaw, J. H., *Intern. Dental J.*, **1**, 48 (1950).
(59) Sognnaes, R. F., *J. Calif. State Dental Assoc.*, **26** (Supplementary Issue), 37 (1950).
(60) Soskin, S., Allweiss, M. D., and Mirsky, I. A., *Arch. Internal Med.*, **56**, 927 (1935).
(61) Soskin, S., and Hyman, M., *Ibid.*, **64**, 1265 (1939).
(62) Soskin, S., and Levine, R., "Carbohydrate Metabolism," Chicago, University of Chicago Press, 1946.
(63) Stefansson, V., *Harper's Magazine*, **171**, 668 (1935); **172**, 46, 178 (1935, 1936).
(64) Stuhlfauth, K., and Prosiegel, R., *Klin. Wochschr.*, **30**, 206 (1952).
(65) Thomas, K., *Arch. Anat. u. Physiol.*, **22** (Suppl.), 249 (1910).
(66) Tolstoi, E., *J. Biol. Chem.*, **83**, 753 (1929).
(67) United Nations Statistical Office, Departments of Economic and Social Affairs, New York, "Demographic Yearbooks, 1949–50 and 1950–51."
(68) U. S. Dept. Agr., "Food and Life, 1938 Yearbook of Agriculture," pp. 152–7, 305, 1939.
(69) Verzar, F., and McDougall, E. J., "Absorption from the Intestine," London, Oxford University Press, 1938.
(70) Wainio, W. W., "The Utilization of Sucrose by the Mammalian Organism," New York, Sugar Research Foundation, *Sci. Rept. Ser.*, **12** (1949).
(71) Wallenfels, K., *Naturwissenschaften*, **38**, 239 (1951).
(72) Weinstein, J. J., *Med. Ann. Dist. Columbia*, **20**, 355 (1951).
(73) White, P. D., "Heart Disease," p. 285, New York, Macmillan Co., 1937.
(74) Woody, E., *Holiday Magazine*, **9** (2), 64 (1951).
(75) Young, L., *Physiol. Rev.*, **19**, 323 (1939).

RECEIVED July 22, 1953.

Sugars in Standardized Foods

BERNARD L. OSER

Food Research Laboratories, Inc., Long Island City, N. Y.

The definitions and standards of identity for foods promulgated under the Federal Food, Drug, and Cosmetic Act designate which sugars may be present as ingredients and in some instances set limits on the amounts or relative proportions. However, no standards have been promulgated for the sugars themselves as a class. In tabulated comparisons among the various types of standardized foods a distinction is drawn between sugars required—i.e., either specified or selected from an optional group—or permitted to be present. Emphasis is placed on the fact that, aside from their sweetening properties, sugars serve numerous technological purposes and in varying degrees. Proposals are discussed for facilitating the standards-making process and for removing unnecessary restrictions on technological progress in the development of new foods and improving the quality, variety, and availability of existing foods.

Prior to the enactment of the Federal Food, Drug, and Cosmetic Act of 1938, it was necessary for enforcement officials in prosecuting cases involving misbranding and adulteration of foods to go to great lengths in establishing to the satisfaction of the courts that a product was in violation of the law. It was necessary to prove the nature and extent of deviation from what was generally conceived to be the commonly accepted food. The norm had to be proved for each case tried, because there were then no legal standards and definitions.

As a guide for officials in the enforcement of the Act of 1906, service and regulatory announcements were published from time to time, including definitions and standards for food products. These were not developed as a result of public hearings, as the present law requires. They had no official standing in the courts. Recognizing the great burden this situation placed upon the Food and Drug Administration, Congress provided in the Food, Drug, and Cosmetic Act of 1938 that "Whenever in the judgment of the Administrator such action will promote honesty and fair dealing in the interest of consumers, he shall promulgate regulations fixing and establishing for any food, under its common or usual name so far as practicable, a reasonable definition and standard of identity, a reasonable standard of quality, and/or reasonable standards of fill of container." Certain foods were exempt from these provisions, among them being fresh or dried fruit with the exception of avocadoes, cantaloupes, citrus fruit, and melons. General regulations governing the administration of the section of the act were then promulgated and, accordingly, the first official standard of identity was adopted in 1939—namely, that for tomato puree and tomato pulp. Subsequently, a considerable number of foods have been standardized and hearings are still being held in the effort to standardize others.

Sugars provide a useful focal point for a consideration of certain aspects of food standardization in general. It is of particular interest at this time to re-examine the effectiveness of standardization as a means of protecting the economic interests of consumers as provided in the administrator's authorization under the law. From the food manufacturers' standpoint it must be recognized that behind every tree in the forest of food standards there hides a specter of misbranding.

The service and regulatory announcements issued prior to the enactment of the present law contained simple definitions for a variety of sugars and related products. These appeared under three major headings as shown in Table I.

Table I. Sugars and Related Substances
(Under service and regulatory announcements)

A. Sugar and sugar products
    Sugar (cane or beet)
    Molasses
    Refiners sirup
    Cane sirup
    Maple sugar
    Maple sirup
    Sorghum sirup

B. Dextrose and related products
    Dextrose, anhydrous
    Dextrose, hydrated
    Glucose
        Mixing glucose
        Confectioner's glucose

C. Honey
    Comb honey
    Extracted honey
    Strained honey

Despite the 15 years which have elapsed since the enactment of the present Food, Drug, and Cosmetic Act, there are no official definitions and standards for sugars as such. [Some standards, grades, and specifications have been issued under the Production and Marketing Act, in official drug compendia (U.S.P. and N.F.), and as Federal Specifications.] However, many foods have been defined which include sugars among their constituents and these sweetening agents are described in more or less detail. This does not mean that the description of corn sirup, for example, in the standard for fruit preserve could serve as the legal basis for an action involving the alleged adulteration of corn sirup per se.

While sugars themselves have never been the subject of Federal Drug Administration standardization hearings, many composite foods which have been standardized include sugars as sweetening agents. In developing the findings of fact upon which the definitions and standards are based, the administrator has described for each standardized food the required or permitted sweetening agents. By combing through all the standards that have been either adopted or proposed as of early 1953, the complete list of sugar products named therein has been assembled in Table II. The findings of fact upon which the standards are based do not contain complete or even adequate specifications for the individual sugars. For example, one does not find any chemical or physical specifications for honey or maple sugar, such as appeared in the old service and regulatory announcements, nor for the varieties of corn sirup available today.

Table II.   Sugars in Standardized Foods
(As of January 1953; under FDC act)

Sucrose (cane or beet)
Invert sugar and sirup
Brown sugar
Invert brown sugar sirup
Refiner's sirup
Molasses
Maple sugar and sirup
Honey

Dextrose
Corn sugar
Corn sirup and solids
Maltose
Malt sirup and solids
Malt cereal extract, solids
Lactose, hydrolyzed

The fact that the administrator has never promulgated standards for sugars and related products may be considered a silent tribute to the industries concerned, as the law provides that food standards are to be promulgated only when his judgment dictates standards to be desirable. Parenthetically, another possible explanation is that the administrator, because of limited staff and facilities, may not have reached this field yet.

This list of sugars is of interest from several points of view. For example, it includes the three major disaccharides—sucrose, maltose, and lactose—in the form of pure sugars. Among the various sugars permitted in both the sirup and dried state is invert sugar, the mixture of dextrose and levulose resulting from the hydrolysis of sucrose, but whereas dextrose alone is permitted in certain foods, levulose is not. This may be due to the fact that levulose is not economically available and hence the hearing records do not contain a proposal for its use; but if this sugar with its high sweetening power should be produced commercially at a suitable price, it would be necessary to reopen hearings in order to obtain permission for its use in standardized foods, as would also be the case for levulose, sorbitol, xylose, or other possible sugars of tomorrow.

Many of the sugars shown in Table II are available commercially in various forms or grades. For example, ordinary sugar or sucrose is sold commercially as liquid sugar; brown sugar can be had in different shades; several varieties of honey are available; and especially significant is the fact that corn sirup, the product of hydrolysis or conversion of cornstarch, is produced with different dextrose equivalents. Nevertheless, the food standards do not concern themselves with these differences, since it is not the sugars that are the subject of the definitions and standards.

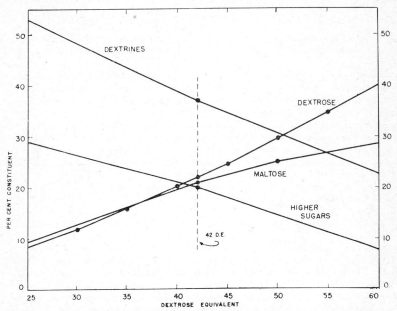

Figure 1. Determination of Carbohydrate Dry Substance in Corn Sirup
Source. Appendix, Corn Industries Research Foundation

Figures 1 and 2 illustrate the variety of products that are commercially possible and available in one category. Here are depicted the changes which occur in corn sirup as the degree of acid hydrolysis or conversion increases; relative concentrations of dextrins and higher sugars diminish, whereas the disaccharide maltose and the monosaccharide dextrose increase in concentration. The dextrose equivalent line of 42 represents the approximate range of conversion of so-called regular corn sirup. Its dextrose equivalent (the total reducing sugar content expressed as dextrose on a dry basis) ranges from 40 to 43. This is essentially the sum of the percentage of dextrose and maltose. Higher sugars, among which have been identified maltotriose and maltotetrose, average close to 20% in regular conversion corn sirup, whereas the dextrins determined by difference constitute approximately 37% of the total solids.

Although the corn products industry does not have any official standards for corn sirup, there have come to be recognized four degrees of conversion. As the dextrose equivalent increases during the progression from low to high conversion sirup, the dextrin solids content diminishes. Associated with these changes are, of course, differences in sweetness, viscosity or consistency, and related physical properties.

In pointing out that Federal Drug Administration standards do not exist for these sugars, it is not intended to imply that there is a necessity for standardization, at least so far as the ultimate consumer is concerned. Control of hydrolysis of cornstarch makes it possible to produce tailor-made sugars in much the same way as the control of hydrogenation of oils has permitted the commercial production of tailor-made fats. It is entirely possible that standardization would place an obstacle in the path of the manufacturer who desires to produce sugars to meet specific needs. Each of the sugars listed in Table II may be used in one or more of the standardized foods, but only one of them—namely, sucrose—is either permitted

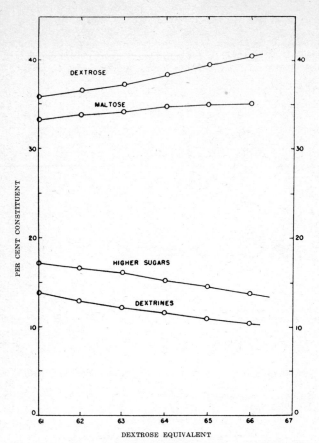

Figure 2. Composition of Acid-Enzyme–Converted Corn Sirup
Dry substance basis
Source. Appendix, Corn Industries Research Foundation

or required in all the standardized foods for which a sweetening agent is specified.

In the definitions of standardized foods it is provided that certain specified ingredients are mandatory, whereas others are optional (Table III). The latter,

Table III.   Ingredients in Standards of Identity

1. Mandatory   M
    Presence is required
2. Optional
    Choice is required   O
    Choice is permitted   P

however, are of two classes, although no official recognition is taken of such a distinction. One class of optional ingredients is comprised of those from which a choice is required to be made; the other class, those from which a choice is permitted but not required. In Tables IV to VI are listed the standardized foods which contain provisions for sweetening agents, the letters M, O, and P indicating whether the use is mandatory, optional, or permissive. The use of a permissive ingredient is a matter of taste or culinary art entirely within the discretion of the food manufacturer. Permitted sugars are often used in combination with or in partial replacement of an optional or mandatory sugar and at specified maximum levels, whereas optional sugars may be used alone.

It will be seen from Table IV that canned fruits may be packed in water, in juice alone, or in one of the specified sirups. These sirups all contain either sucrose or invert sirup as one of the sweetening agents. Sucrose, of course, is normally inverted on standing in acid fruits or beverages. Dextrose or corn sirup (liquid or dried) is permitted to be used, but only to a limited extent in terms of the total solids content. In canned vegetables sucrose, and in some cases dextrose, may be used. Neither of them is required.

Table V shows that in jellies, jams, and preserves a choice must be made between sucrose and invert sugar sirup as one of the required sweetening agents.

Table IV. Canned Fruits and Vegetables

Sugar	Peaches	Apricots	Cherries	Pears	Pine-apple[a]	Fruit Cocktail	Peas	Corn	Vege-tables
Sucrose	O[b]	O[b]	O[b]	O[b]	P	O[b]	P	P	P[c]
Invert sugar									
Invert sugar sirup	O[b]	O[b]	O[b]	O[b]	P	O[b]			
Brown sugar									
Invert brown sugar sirup									
Molasses									
Refiner's sirup									
Dextrose	P[d]	P[d]	P[d]	P[d]		P[d]	P		P[c]
Corn sugar									
Corn sirup, dried	P[d]	P[d]	P[d]	P[d]					
Corn sirup	P[d]	P[d]	P[d]	P[d]		P[d]			
Glucose sirup, dried									
Glucose sirup									
Maltose									
Malt cereal extract, dried									
Malt sirup, dried									
Malt sirup									
Maple sugar									
Maple sirup									
Honey									
Lactose, hydrolyzed									

[a] Proposed.
[b] Except if water- or juice-packed.
[c] Except canned mushrooms.
[d] In combination with sucrose or invert sugar sirup.

Table V. Fruit Products and Frozen Desserts

Sugar	Fruit Butter	Fruit Jelly	Preserves, Jams	Ice Cream[a]	Sherbet, Water Ices[a]	Frozen Fruits[a]
Sucrose	O[b]	O	O	O	O	P
Invert sugar						P
Invert sugar sirup	O[b]	O	O	O	O	P
Brown sugar	O[b]			O		
Invert brown sugar sirup	O[b]					
Molasses				O		
Refiner's sirup						
Dextrose	P[c]	P[d]	P[d]	O	O	P[e]
Corn sugar						
Corn sirup, dried	P[c]	P[d]	P[d]	O	O	P
Corn sirup	P[c]	P[d]	P[d]	O	O	P
Glucose sirup, dried						
Glucose sirup	P[c]	P[d]	P[d]			P
Maltose						
Malt cereal extract, dried						
Malt sirup, dried				O		
Malt sirup				O		
Maple sugar				O		
Maple sirup				O		
Honey	O[b, c]	O[d]	O[d]	O		
Lactose, hydrolyzed						

[a] Proposed.
[b] Not required if fruit juice is used.
[c] In combination with sucrose, invert sugar sirup, brown sugar, or invert brown sugar sirup.
[d] In combination with sucrose or invert sugar sirup except honey which may also be used alone.
[e] In combination with sucrose or invert sugar sirup.

In the case of fruit butters this is extended to include brown sugar or invert brown sugar sirup. Honey may also be used in these foods either alone or in combination with sugar or invert sirup. However, corn sirup or dextrose may be used only to a limited extent in combination with these sugars.

The standards for frozen fruits and frozen desserts have not yet been finally promulgated. However, it may be seen from Table V that whereas various sweetening agents may be used in frozen fruits, none is required. In the case of ice cream an even wider option is permitted, but the use of some sweetening agent is essential to its identity.

From Tables IV and V it can be seen that in all sweetened fruit products sucrose or invert sugar sirup must be used in whole or in part, except in certain cases where honey is used as the sole sweetening agent. (Fruits may be packed in water or in fruit juice.) The extent to which dextrose or corn sirup products are permitted in partial replacement of sucrose has been limited for each of the standardized foods. To some observers these limitations appear to be based on technological or organoleptic considerations rather than on economic grounds. For example, where sweetness is an important factor, as in preserves or jellies, it appears that the sole basis for fixing limitations is to ensure the normally expected degree of sweetness. Assuming this requirement to be satisfied, it is difficult to understand how variations in sweetness promote honesty and fair dealing in the interests of consumers.

The remaining classes of standardized foods for which sweetening agents are prescribed are shown in Table VI. In the case of bread, the addition of a sugar ingredient is not mandatory. The function of sugar in this class of products is not exclusively that of a sweetening agent. Bread can still be identified as bread, whether or not sugars are used in its composition. A similar situation prevails with respect to cheese spreads, catsup, mayonnaise, and salad dressings. However, even where sweetening agents are only permissive, the use of a sugar ingredient other than from the prescribed list would be in violation of the standard. Thus under the respective definitions of these foods, dextrose may be used in catsup, but dried corn sirup may not (amended October 9, 1954, to permit the use of dried corn sirup); brown sugar is permitted in bread, but maple sugar is not; sucrose must be used in sweetened condensed milk, but invert sugar may not.

From the number and variety of sugars whose use is permitted in standardized

Table VI. Miscellaneous Foods

Sugar	Bread and Rolls	Sweet Condensed Milk	Cheese Spreads	Sweet Chocolate	Milk Chocolate	Catsup	Mayonnaise, French, and Salad Dressings
Sucrose	P	M	P	M	M	P	P
Invert sugar	P						P
Invert sugar sirup	P						P
Brown sugar	P[a]			P[b]	P[b]		
Invert brown sugar sirup							
Molasses	P			P[b]	P[b]		
Refiner's sirup	P						
Dextrose	P	P[c]	P	P[c]	P[c]	P[c]	P
Corn sugar			P				
Corn sirup, dried	P		P	P[c]	P[c]		
Corn sirup		O[d]	P				P
Glucose sirup, dried							P
Glucose sirup	P						
Maltose	P		P				
Malt cereal extract, dried	P			P[b]	P[b]		P
Malt sirup, dried	P						P
Malt sirup	P		P				
Maple sugar				P[b]	P[b]		
Maple sirup							P
Honey	P			P[b]	P[b]		
Lactose, hydrolyzed			P				

[a] Light colored.
[b] For flavor rather than sweetening.
[c] In combination with sucrose or invert sugar sirup.
[d] Presence must be indicated in name.

foods it must be obvious that factors other than mere sweetness enter into their choice. It is reasonable to assume that the manufacturer will select the most economical and available sugar or combination of sugars that satisfy the organoleptic, physical, chemical, and technological requirements for any given product. Aside from considerations of cost and supply, Table VII lists the major determining factors from the technological standpoint. Each of these items cannot be analyzed here. However, the large variety of sugars available today, in different degrees of hydrolysis or conversion, dehydration, granulation, or crystallization, permits the food technologist to exercise considerable discrimination in respect to choice of sweeteners.

Table VII.   Factors Influencing Choice of Sugars

Organoleptic	Sweetness, flavor, capacity to blend or bring out taste (fruit products)
Solubility	Quick sweetness; high concentration (preservation)
Osmotic pressure	Preservation against microorganism. Transfer into fruit tissue
Density	Body, consistency, viscosity (fruit, sirups, confections)
Crystallization	Stability on exposure (sugar bloom); grain or texture (confections) chewiness
Hygroscopicity	Moisture retention (creams, icings)
Freezing point depression	Ice cream, sherbet
Fermentability	Aid to leavening
Caramelization	Browning of crust
Preservation	Against microbiological or chemical spoilage
Nutrition	Caloric value

These considerations also serve to emphasize how long drawn out is the process of developing the basic information necessary for the promulgation of food standards. Public hearings as provided by law place a great burden on the government and on industry. It is even possible that smaller segments of industry or of the consuming public may not be able to afford the time and expense of adequate representation at these hearings.

It would seem desirable therefore to explore whether standardization of foods ought not to be strictly limited to those situations where it is the only effective means of providing the necessary protection of the consumer's economic interests and to consider the possibility that more realistic and comprehensive label declarations of ingredients, including the declaration of mandatory as well as optional ingredients, would accomplish this purpose equally well.

The findings of fact upon which definitions and standards of identity are based are derived from the record established at the public hearings. It might help shorten these hearings to avoid taking repetitious testimony on the varieties of sugars to be used in different foods. This could be accomplished by adopting the definitions and standards for all permitted sugars, allowing the food manufacturer greater freedom to choose sugars best adapted to his product so long as they were selected from the approved list. If, in special cases, the economic interests of consumers require that the specific sugars used be declared on the label of any particular food, such provision would be made in the definition and standard for that food. Ample precedent exists for permitting a choice of suitable, harmless ingredients without actually declaring their presence. To require that all sugars be identified on the label would be unnecessary and unreasonable. For example, an ancient prejudice which has no foundation in modern science or technology relegates beet sugar to a position somewhat inferior to cane sugar, and corn sugar to a position beneath either. However, consumer education is a continuing process and the food industry must accept this challenge. The prejudices against canned foods, pasteurized milk, white flour, and certain sugar products that are now recognized as wholesome and nutritious foods did not die easily.

The promulgation of food standards is a powerful and useful function in the administration of the Food, Drug, and Cosmetic Act. This authority has been recognized by the highest courts in the land. The wisdom and legality of food standardization can no longer be questioned. It is well to keep in mind, however, that its function is to promote honesty and fair dealing in the interest of consumers. To the extent that it is compatible with this function, food standards should not impose unnecessary burdens on industry or impede technological progress in developing new foods or in augmenting the quality, variety, or availability of existing foods.

That the present standard-making procedure is in need of streamlining is indicated by the appointment of a special committee of the Food and Nutrition Board of the National Research Council, under the chairmanship of R. R. Williams, which is studying the situation. It is believed that it would promote technological advancement to adopt regulations permitting a limited period for trial use of new ingredients in standardized foods, prior to actual reopening of standards hearings. Members of the bar specializing in food and drug law have proposed legislation (H.R. 5055, subsequently introduced into the 83rd Congress as H.R. 6434 and enacted as Public Law 335. This permits issuance, amendment, or repeal of standards upon petition of any interested party without the necessity of holding public hearings except upon the filing of objections to the proposed order.) to expedite standards hearings by providing for preliminary informal agreement on noncontroversial subjects, leaving for the hearings only those items about which questions are raised. The food industry will look with approval upon such advances on the legislative front.

It has not been the writer's purpose to be critical of the manner of enforcement of the standards-making provisions of the Food, Drug, and Cosmetic Act. The Food and Drug Administration with its limited personnel has done an excellent job in promulgating standards under the law and regulations as they now exist. However, it is obvious that standards are growing rapidly, inconsistencies have arisen, and the hearings have, in some instances at least, become unduly prolonged by the consideration of issues that are, strictly speaking, outside the scope of "honesty and fair dealing in the interests of consumers." One of these issues—namely, safety of food ingredients—will undoubtedly be subject to consideration outside of standards hearings if legislation now under consideration should be adopted.

RECEIVED April 15, 1953.

INDEX

Page numbers in bold-face type designate authors or main topics of papers in this symposium. Only the first page of the paper is so designated; repetitions of the key word on later pages in that paper are not indexed.

T

U

V

W

X

Y

Z